Ozone Risk Communication and Management

Edward J. Calabrese
Charles E. Gilbert
Barbara D. Beck
Editors

LEWIS PUBLISHERS

Library of Congress Cataloging-in-Publication Data

Ozone risk communication and management/edited by Edward J. Calabrese.

   p. cm.
   Bibliography: p.
   Includes index.
   1. Atmospheric ozone—Environmental aspects—United States—Congresses.
2. Atmospheric ozone—Hygienic aspects—United States—Congresses. 3. Health
risk assessment—United States—Congresses. I. Calabrese, Edward J.

TD885.5.O85096   1990   88-11722
363.7—dc19
ISBN 0-87371-130-0

LEWIS PUBLISHERS, INC.
121 South Main Street, Chelsea, Michigan 48118

PRINTED IN THE UNITED STATES OF AMERICA

Edward J. Calabrese is a board-certified toxicologist who is professor of toxicology at the University of Massachusetts School of Public Health, Amherst. Dr. Calabrese has researched extensively in the area of host factors affecting susceptibility to pollutants. He is the author of more than 240 papers in refereed journals and 10 books, including *Principles of Animal Extrapolation, Nutrition and Environmental Health Vols. I and II, Ecogenetics,* and others. He has been a member of the U.S. National Academy of Sciences and NATO Countries Safe Drinking Water committees and, most recently, has been appointed to the Board of Scientific Counselors for the Agency for Toxic Substances and Disease Registry (ATSDR).

Dr. Calabrese was instrumental in the conceptualization and development of the Northeast Regional Environmental Public Health Center and was appointed its first director. The center's mission includes communication, education, and research.

Charles E. Gilbert is a research associate in toxicology at the University of Massachusetts School of Public Health, Amherst. Dr. Gilbert received his BS, MSc, and PhD degrees from the University of Massachusetts.

He was the assistant director for the Childhood Lead Poisoning Prevention Program, Massachusetts Department of Public Health, where he directed improvements in environmental management, case management, and education programs. His research interests are in the area of factors that affect human susceptibility to biological, chemical, and physical agents and how these affect health.

Charles Gilbert worked in the development of the Northeast Regional Environmental Public Health Center and was appointed its first assistant director. This center is a cooperative organization of the New England Public Health Departments and the University of Massachusetts School of Public Health.

Barbara D. Beck is an expert in toxicology and in health risk assessment for environmental chemicals and the author of over 20 book chapters and journal articles on these topics as well as on variations in susceptibility to environmental pollutants. She is associated with Gradient Corporation, Cambridge, Massachusetts, where her projects include: litigation support for a case involving dioxin contaminated soil; evaluation of health significance of clean-up levels for As and Pb at waste sites; and review and evaluation of scientific literature on toxicity of components of automobile exhaust. Before joining Gradient, Dr. Beck was Chief of the Air Toxics Staff in Environmental Protection Agency Region I, where she performed risk assessments for air and soil pollutants and supervised a staff with responsibility in air toxics and radiation. Dr. Beck was also one of 10 members of EPA's national Risk Assessment Forum. Prior to that she was a Research Associate in Environmental Science and Physiology and a Fellow in the Interdisciplinary Programs in Health at the Harvard School of Public Health, where she developed an assay for lung toxicity and authored several chapters of a monograph on susceptibility to inhaled pollutants. She is at present a Lecturer in Toxicology at Harvard and a member of the Science Advisory Panel to the Maine Bureau of Health.

Preface

Current levels of a majority of the air pollutants monitored by the U.S. Environmental Protection Agency have decreased, indicating that for the past 10 years clean air programs have been, at least in part, successful. Reductions in atmospheric concentrations of carbon monoxide, sulfur dioxide, nitrogen dioxide, and lead are all examples of these successes. Ozone, however, has not been as responsive to pollution control programs because of the inherent complexity in its formulation and decomposition. *Ozone Risk Communication and Management* provides the scientific and technical reasons for the limited success in controlling ozone pollution and offers insightful recommendations on how to achieve public health goals.

The major topics discussed in the text are: generation and transport of ozone; health effects; effects on vegetation; effects on materials; and implications for a National Ambient Air Quality Standard. Finally, the book discusses various approaches for reducing ozone levels.

Chapters 1, 2, and 3 discuss the chemical and meteorological aspects of photochemical oxidants.

In the first chapter Dr. Demerjian notes that considerable advances have been made in understanding the complex photochemical smog generation processes. These advances include improved mechanisms for categorizing the photochemical oxidant cycle, enhanced understanding of urban boundary processes, evaluation of oxidant models, and the evaluation of analysis accuracy. Dr. Demerjian discusses the importance of understanding and recognizing the limits of model predictability and how these limitations should be incorporated into public policy. In Chapter 2, Dr. Sham describes the mechanisms governing ozone transport, and factors that modulate such transport processes. In Chapter 3, Mr. Burkhart explains how meteorological factors make accurate interpretation of ozone trends over time difficult, and their impact on the effectiveness of implementation strategies.

Photochemical oxidant impacts on the environment, such as crops and forests, and commercial goods such as vehicle tires are considered in Chapters 4, 5, and 6.

In Chapter 4, Dr. Manning describes a variety of species of plants that are very sensitive to ozone pollution. These include plants that are of importance for agriculture, such as spinach and tobacco. Forest susceptibility to ozone pollution, particularly the pine tree, is presented by Drs. Adams and Taylor in Chapter 5. As Dr. Manning, Dr. Adams, and Dr. Taylor point out, some plants are susceptible to ozone even at levels less than 0.08 ppm.

It is also clear that ozone, a reactive oxidizing gas, affects nonbiological receptors. It has been known for two decades that ozone is a particularly potent oxidizing agent, and can affect rubber and other synthetic elastomers at relatively low atmospheric concentrations. There are also effects on textile fibers, such as fading of dyes. Some of these effects appear to be compounded by exposure to other pollutants such as nitrogen oxides. In Chapter 6, Dr. Kuczkowski describes the effects of ozone on tires and explains how these effects are mitigated.

As presented in Chapters 7, 8, 9, and 10, toxicological experiments, epidemiological evaluations, and clinical studies have shown that exposure to ozone results in a variety of health effects. Animal experiments prove that even at ambient levels of ozone, morphological changes occur at the alveolar level of the lung. Ozone toxicity in animal models is described by Dr. Barry in Chapter 7. At very low levels of ozone, rodents experience increased susceptibility to infection. Dr. McDonnell, in Chapter 8, discusses clinical studies which are likely to be the most directly relevant to humans. In these studies, reversible decrements in lung function and symptoms such as a cough or pain upon inhalation have been reported. Extrapulmonary changes are also observed with ozone exposure, and these are described in Chapter 9 by Drs. Canada and Calabrese. Epidemiological studies suggest that ozone exposure may be responsible for adverse health effects in general population. Epidemiological studies on short term health effects associated with ozone are discussed by Drs. Dockery and Kriebel in Chapter 10.

One issue of priority consideration in the setting of National Ambient Air Quality Standards is that of the susceptible population. Dr. McKee points out in Chapter 11 that the primary ozone standard must be set to protect the majority of individuals within the susceptible population from a defined health effect. The definition of an *ozone* susceptible person is more complex than with *lead,* where the susceptible population is children, who are easily defined and quantified. The wide range of ozone susceptible populations includes exercisers and individuals with preexisting lung disease. There are also variations in individual responsiveness to ozone, and some may be considered "hyper-responders." While there are some estimates on the fraction of the population that may be considered hyper-responders, the group cannot be adequately defined without more information, particularly regarding physiological mechanisms.

In the final chapter (Chapter 12) John Hanisch addresses the EPA's strategies for reduction of ozone levels, including issues of technical, economic feasibility and the likelihood for support and social acceptance by the general population.

Edward J. Calabrese
Charles E. Gilbert
Amherst, MA

Barbara D. Beck
Cambridge, MA

Contents

CHAPTER 1

Factors Affecting the Formation of Ozone

Kenneth L. Demerjian

INTRODUCTION

Ozone is an atmospheric pollutant of significant consequence to public health and welfare. Even prior to the establishment of the National Ambient Air Quality Standards under the 1970 Clean Air Act, it had been identified as an important factor in the smog episodes that plagued urban areas having hot and sunny climates, prone to meteorological air stagnations, and with high motor vehicle emission densities. In addition, recent advances in the knowledge of the importance of ozone in controlling the chemical composition and climate of the troposphere has highlighted the necessity that the factors affecting its production must be thoroughly understood.

Ozone, unlike the other criteria pollutants, does not have a primary anthropogenic emission source, but rather is produced by secondary reactions resulting from anthropogenic precursor emissions (predominantly hydrocarbons and oxides of nitrogen). Ozone is also unique in that it has a natural component which, under certain atmospheric conditions, can contribute to the ozone concentration a significant fraction of the ambient air quality standard.

The factors affecting ozone formation in the atmosphere are reviewed starting with a brief discussion of the origins and levels of background ozone, followed by a review of the chemical mechanism development process and the chemistry of ozone production. The chapter then addresses the application of chemical mechanisms in developing quantitative relationships between precursor emissions

1

and ozone production and their use and limitation in the control strategy development process, and finally ends with a brief discussion of the ozone transport issue.

BACKGROUND OZONE

Ozone in the clean troposphere is the result of stratospheric-tropospheric exchange as well as production from chemical reactions involving the oxidation of CO, CH_4, and possibly other biogenic hydrocarbons. The ozone that results from these two sources may be destroyed by chemical reactions as well as by interactions with the earth's surface. The net production of ozone in the clean tropospheric environment from chemical oxidation reactions involving CH_4 and CO depends upon the availability of NO_x. Current chemical mechanisms of the clean troposphere indicate that NO_x levels greater than 30 ppt are sufficient to maintain net ozone production. The background levels of ozone resulting from these two natural sources range between 20 and 50 ppb, and have both a seasonal and latitudinal component.[1] The CH_4-CO-NO_x reaction cycle cannot explain ozone exceedances observed in polluted atmospheres.

In polluted atmospheres, unlike the clean troposphere, the rate of ozone formation is not necessarily proportional to the concentration of NO_x, but varies in a complex way that is dependent upon the ratio and concentrations of the hydrocarbons and NO_x as well as the chemical composition of the hydrocarbons themselves. The chemistry of ozone in the polluted troposphere has been actively studied for over 30 years. The research emphasis has been predominantly in elucidating the processes for oxidant/ozone formation in urban areas. The focus of interest in urban areas is mainly due to the high emission densities of ozone precursors (hydrocarbons and NO_x) which typically result in very high ozone concentrations under summertime meteorological conditions. In many cases the concentration levels of ozone in urban areas exceed ambient air quality standards which have been set to protect public health and welfare.

Several reviews of polluted atmospheric chemistry are available,[2-4] as are detailed discussions of reaction mechanisms.[5-9] This discussion provides a brief overview of the chemistry of the polluted troposphere and the factors affecting the formation of ozone. It is suggested that those readers interested in a more detailed account consult the cited references for further information.

THE CHEMICAL MECHANISM DEVELOPMENT PROCESS

The development of chemical mechanisms for the purpose of representing the transformation of pollutants and background trace constituents in the atmosphere has evolved over approximately a 20 year period and has considered several principal components in the development process. The components that form the basis for the methodology of this development process are illustrated in Figure 1.

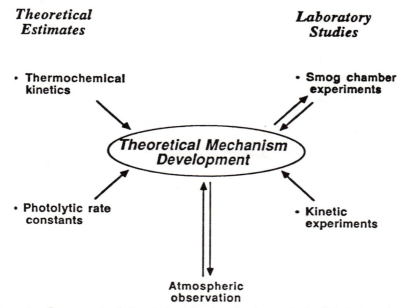

Figure 1.   Components of chemical mechanism development of polluted atmosphere.

Historically, each of these components has assumed a dominant role in the mechanism development process, somewhat reflecting the state of the science during the various development stages. For example, in the early 1970s at the onset of research and development activities in chemical mechanisms for the simulation of atmospheric transformations in polluted environments,[5,11] a majority of the elementary reaction steps were generally theorized from thermochemical kinetic estimates based on the methods introduced by Benson in 1968.[12] Mechanisms were developed using data from smog chamber experiments as a basis set for truth; that is, mechanisms were judged on their success in fitting the concentration-time profiles of the experimental data. The assumption was that the smog chamber experiments provided a representative analogue of the chemical systems operative in real-world polluted atmospheres, therefore allowing the extrapolation of the developed mechanisms to simulating the chemical transformations in polluted atmospheres. Figure 2 provides a typical example of concentration-time profiles from irradiations of a propene-NO_x system in a smog chamber.[13]

Many critical elementary reaction steps were identified in the process described above for which no laboratory chemical kinetic data existed. The importance of these reactions in understanding the mechanistic transformations of pollutant species in the atmosphere created a forcing function which stimulated laboratory chemical kinetic studies. Rate parameters and elementary modes of reaction for a large variety of species and reactions were provided by the chemical kinetic

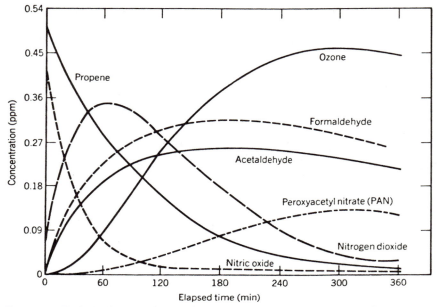

Figure 2. Typical primary and secondary pollutant profiles in a propene-NOx irradiation in a smog chamber.

community which introduced significant advances and refinements in mechanisms in the late '70s and '80s. Recent reaction rate constant reviews[14-16] illustrate the progress that has been made in this important area.

As the feedback process above was occurring, so also was a process between the model development and smog chamber communities. The mechanistic modelers attempted from the start to develop and test their mechanisms for as many chamber systems and data sets as possible. In doing so the modeling community recognized certain limitations in the databases and chamber systems they were utilizing, and initiated some guidelines for the smog chamber experimentalists. This resulted in additional enhancements in the mechanism development, and represents an important methodological component in the evolutionary process.

THE CHEMISTRY OF OZONE PRODUCTION

The chemistry of ozone production occurs in sunlight-irradiated polluted atmospheres and involves the interaction of a host of chemical species. These include: hydrocarbons such as alkanes, alkenes, and aromatics; organics such as aldehydes and ketones; atomic oxygen [O_3P] and its first electronic excited state [O^1D]; ozone (O_3); nitrogen dioxide (NO_2); nitric oxide (NO); nitrogen trioxide (NO_3); dinitrogen pentoxide (N_2O_5); hydroxyl radical (HO); hydroperoxyl radical (HO_2); alkylperoxyl radicals (RO_2); acylperoxyl radicals ($RC(O)O_2$); nitric acid (HNO_3); and peroxyacetyl nitrate (PAN), to name a few of the more important species.

Chemical pathways are a very important element of a complex interactive system of processes that govern the fate of chemical constituents in the atmosphere and their ultimate disposition. It is important to recognize that our understanding of the photochemical oxidant phenomenon can only be accomplished through an integrated perspective of all the operational pathways that affect their formation, fate, and transport in the environment. Figure 3 provides a simplified schematic

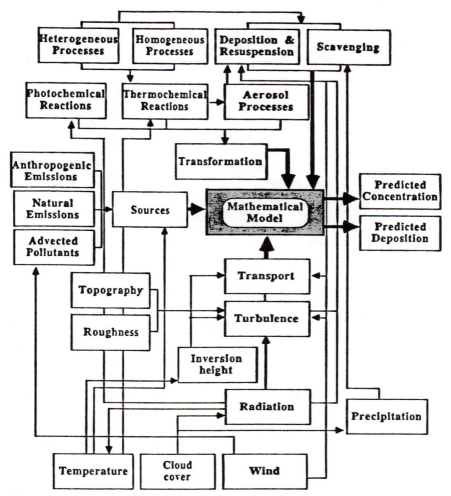

Figure 3.   A schematic of the major process components contributing to photochemical oxidant models.

illustration of some of the more important physical and chemical pathways thought to be of significance with regard to photochemical oxidants and their treatment in mathematical models.

The chemical pathways leading to the conversion of nitrogen oxides and organic compounds to form photochemical oxidants in the atmosphere, of which ozone is the most dominant species, occur through a variety of oxidizing agents formed via a complex sequence of gas phase reactions within the atmosphere.[2-9] Oxidizing species, such as molecular ozone, hydrogen peroxide, organic peroxides, peroxyacetic acid, and free radicals such as HO, RO, HO_2, RO_2, and NO_3 have been implicated as important contributors to ozone chemistry and the mechanisms of their formation has been substantially characterized.[4-11] In the formulas which follow, R represents a methyl (CH_3), ethyl (C_2H_5), or other more complex hydrocarbon fragment. The paths by which these species are formed and decomposed are important elements in understanding the chemical changes that occur in the polluted atmosphere.

PHOTOCHEMICAL REACTIONS AND OXIDATION CYCLES

The major portion of the total oxides of nitrogen emitted by combustion sources into the atmosphere is in the form of nitric oxide (NO). The rate at which NO is converted to nitrogen dioxide (NO_2) through oxidation by molecular oxygen in air is not sufficient to explain the high conversion rates of nitric oxide to nitrogen dioxide observed under ambient concentrations in relatively clean or polluted atmospheres. The key to the observed nitric oxide to nitrogen dioxide conversion lies in a cyclical sequence of reactions involving transient free radical species and reactive hydrocarbon and organic compounds present in the atmosphere. Within this same sequence of reactions lies the key to ozone formation in the atmosphere.

A very important element of the atmospheric photooxidation process is knowledge of those photochemical constituents which absorb sunlight and form active species which decompose or react with other constituents in the atmosphere. Nitrogen dioxide, a dominant sunlight absorber, photodissociates upon absorbing wavelengths of light < 430 nm. This photolytic reaction results in the formation of $O(^3p)$ atom, and a molecule of nitric oxide. The efficiency of this reaction as well as all others to be discussed is wavelength dependent. In the case of NO_2 photodissociation, its absorption spectrum is relatively flat over the available solar actinic flux, but its quantum yield to form dissociative products drops dramatically from approximately 1.0 at wavelengths < 390 nm to 0 at wavelength > 420 nm.

Ozone is produced photochemically in the atmosphere as a result of the photolysis of NO_2 which forms $O(^3P)$ followed by its reaction with molecular oxygen.

$$NO_2 + hv \ (<430 \text{ nm}) \longrightarrow O(^3P) + NO \qquad (1)$$

$$O(^3P) + O_2 \ (+M) \longrightarrow O_3 + (+M) \qquad (2)$$

Depletion of ozone can occur in part by its reaction with NO resulting in the reformation of NO_2,

$$NO + O_3 \longrightarrow NO_2 + O_2 \qquad (3)$$

and through its photolysis and subsequent reaction of its fragmented product $O(^1D)$.

$$O_3 + hv \ (<306 \text{ nm}) \longrightarrow O(^1D) + O_2 \qquad (4)$$

$$O(^1D) + M \longrightarrow O(^3P) + M \qquad (5)$$

$$O(^1D) + H_2O \longrightarrow 2HO \qquad (6)$$

Reactions 4 and 6 are an important source of hydroxy radicals (HO) in the relatively clean atmosphere. Hydroxy radicals are a key constituent in the chemical reaction cycles affecting oxidation of the trace gases and the formation of acidic species.

A pseudo chemical steady state relationship involving reactions (1)–(3) has been shown to be a reasonably valid approximation for ozone concentrations in the atmosphere during the day for a considerable range of atmospheric pollutant conditions.[17-18] The relationship:

$$[O_3] \approx [NO_2] \, J_1 \, / \, [NO] \, k_3 \qquad (I)$$

where J_1, the first order photodissociation rate constant for reaction (1), is proportional to the sunlight intensity in the atmosphere over the available wavelength range (290—490 nm) that NO_2 dissociates. The diurnal and seasonal variation of sunlight intensity is therefore an important component in ozone production and the overall atmospheric oxidation cycle.[15-19]

Another source of hydroxy radicals is through the photolysis of nitrous acid:

$$HONO + hv \ (<400 \text{ nm}) \longrightarrow HO + NO \qquad (7)$$

Nitrous acid is formed via the reaction of NO, NO_2, and water, which is thought to be significantly surface catalyzed and can revert back to the original reactants through its bimolecular dissociation.

$$NO + NO_2 + H_2O \longrightarrow 2HONO \qquad (8)$$

$$HONO + HONO \longrightarrow NO + NO_2 + H_2O \qquad (9)$$

The significance of this source for free radical initiation in the early stages of the photochemical oxidation cycle due to suspected buildups in nitrous acid within shallow, moist nocturnal boundary layers rich in NO and NO_2 remains to be substantiated.[20-22]

Hydrocarbons/organics present in the atmosphere,[23-28] introduce a large number of new reactions to consider. The most important of these reactions involve the oxidation of the hydrocarbons, initiated by their reaction with hydroxy radical.

$$HO + Hydrocarbon \longrightarrow R + H_2O \tag{10}$$

Hydrocarbon in the reaction above refers to alkanes, alkenes, aromatics or any organic compound having C-H bonds. The organic free radical (R) produced as a result of the HO attack on the hydrocarbon contains a free electron which can react with an oxygen molecule in the air quite rapidly to form an organic peroxy radical (RO_2)

$$R + O_2 \longrightarrow RO_2 \tag{11}$$

The organic peroxy radical typically reacts with NO to form NO_2 and an organic oxy radical (RO)

$$RO_2 + NO \longrightarrow RO + NO_2 \tag{12}$$

Hydrogen abstraction from RO by molecular oxygen will produce a hydroperoxy radical (HO_2) and in most cases a carbonyl compound $(R(C{=}O)H)$, which may be aldehydes, such as formaldehyde and acetaldehyde; dicarbonyls, such as glyoxal and methyl glyoxal; and ketones, such as methyl ethyl ketone and hydroxy methyl ketone.

$$RO + O_2 \longrightarrow R(C{=}O)H + HO_2 \tag{13}$$

The aldehydes formed in the RO oxidation, and which are also emitted from many combustion sources, may photolyze or react with HO, introducing another important source of radicals in the atmosphere

$$R(C{=}O)H + hv\ (<400\ nm) \longrightarrow R + H(C{=}O) \tag{14}$$

$$R(C{=}O)H + HO \longrightarrow R(C{=}O) + H_2O \tag{15}$$

The hydroperoxy radical, produced in the oxidation, can then react with NO to form NO_2 and another HO radical, thereby completing a reaction chain cycle.

$$HO_2 + NO \longrightarrow HO + NO_2 \tag{16}$$

The interaction of organic free radicals produced by hydrocarbon oxidation with NO and NO_2 represents an important aspect of the chemistry of the oxides of nitrogen in the ambient atmosphere. The reactions of particular significance are those which are radical and NO_x sinks or which result in temporary storage of these species. The reactions of HO with NO_2 and NO will be discussed in a later section with regard to their acid forming potential.

Peroxynitric acid is formed by the reaction of HO_2 with NO_2 and thermally decomposes back to the original reactants.[29]

$$HO_2 + NO_2 + M \longrightarrow HO_2NO_2 + M \qquad (17)$$

$$HO_2NO_2 \longrightarrow HO_2 + NO_2 \qquad (18)$$

At temperatures prevalent during summertime conditions, it appears that peroxynitric acid does not represent an appreciable sink for NO_2, due to its rapid thermal decomposition. At lower temperatures (e.g., winter or upper tropospheric conditions) HO_2NO_2 can achieve higher concentrations and may be an appreciable sink for NO_2.

The reactions of RO, RO_2, and $R(C{=}O)O_2$ with NO and NO_2 are key processes in the conversion of NO to NO_2 and the formation of organic nitrites and nitrates. RO radicals which predominantly react with O_2 in the atmosphere may to a much lesser extent react with NO and NO_2.

$$RO + NO \longrightarrow RONO \qquad (19)$$

$$RO + NO_2 \longrightarrow RONO_2 \qquad (20)$$

The reaction of organic peroxy radicals with NO is generally thought to proceed by the oxidation of NO to NO_2 with the formation of an RO radical. It has recently been postulated that longer chain peroxyalkyl radicals (carbon number >4) formed, for example, during alkane photooxidation will add NO to form an excited complex that can stabilize to produce an organic nitrate.[30]

$$RO_2 + NO \longrightarrow [RO_2NO]^1 \longrightarrow RONO_2 \qquad (21)$$

The formation of organic peroxy nitrates (RO_2NO_2), the analogue to peroxynitric acid, is also suspected to be in thermal equilibrium with its reactants, though measured rate constants are currently unavailable.

$$RO_2 + NO_2 \longrightarrow RO_2NO_2 \qquad (22)$$

$$RO_2NO_2 \longrightarrow RO_2 + NO_2 \qquad (23)$$

Peroxyacyl nitrates, and in particular peroxyacetyl nitrate (PAN) which is the most commonly observed organic nitrate in clean and polluted atmospheres,[31-34] exist in thermal equilibrium with the peroxyacyl radical and NO_2

$$R(C=O)O_2 + NO_2 \longrightarrow R(C=O)O_2NO_2 \tag{24}$$

$$R(C=O)O_2NO_2 \longrightarrow R(C=O)O_2 + NO_2 \tag{25}$$

A competition for the peroxyacyl radical exists due to its reaction with NO discussed previously. PAN chemistry is an important feature of the hydrocarbon—NO_x photooxidation cycle in the troposphere and may play an exceedingly important role as a temporary sink for radicals and NO_x[35-36] in affecting the long distance transport of these species and their contribution to the continental background.

Both NO and NO_2 react with HO to form their respective acids, nitrous and nitric acid.

$$HO + NO + M \longrightarrow HONO + M \tag{26}$$

$$HO + NO_2 + M \longrightarrow HONO_2 + M \tag{27}$$

In the case of nitrous acid, its relatively fast photolysis to return the reactants HO and NO results in its being a rather insignificant source of acid, during the daylight hours. As mentioned earlier, its significance as a source of acid in the moist nocturnal boundary layer remains to be substantiated. Nitric acid formation, on the other hand, via reaction,[27] represents one of the major atmospheric sinks for the highly reactive HO radical and NO_2 and a primary source of this acid species.

An additional source of atmospheric nitric acid is through the chemistry associated with the transient species, nitrogen trioxide NO_3. Nitrogen trioxide is formed predominantly from the reaction of nitrogen dioxide and ozone. This reaction, though much less competitive than the ozone-nitric oxide reaction, can be significant in the latter stages of the photooxidation cycle.

$$O_3 + NO_2 \longrightarrow NO_3 + O_2 \tag{28}$$

The NO_3 can enter into a reversible reaction with NO_2 to form dinitrogen pentoxide N_2O_5, the reactive anhydride of nitric acid.

$$NO_3 + NO_2 \longrightarrow N_2O_5 \tag{29}$$

$$N_2O_5 \longrightarrow NO_3 + NO_2 \tag{30}$$

The N_2O_5, which is in temperature dependent equilibrium with NO_2 and NO_3, can react with water to form nitric acid, $HONO_2$.

$$N_2O_5 + H_2O \longrightarrow 2HONO_2 \tag{31}$$

The rate constant for this reaction is sufficiently low, so that it is not considered an important source of nitric acid formation during the photochemically active daylight period, but there is some indication that this reaction process may be a significant source of NO_x to $HONO_2$ conversion during the nighttime period, when NO_2, O_3, and water are all available and not being dominated by other competing reactions.[38-40]

Recent studies of the photochemistry of NO_3, which absorbs strongly in the red, have confirmed that its photochemical lifetime is extremely short (of the order of 10 s) during the daytime. This result, and the very fast reaction of NO_3 with NO, reaction (32), (which is always present during the day in photochemical equilibrium with NO_2 and ozone) leads to extremely low daytime concentrations of NO_3 (and thus N_2O_5) in the sunlit troposphere.

$$NO_3 + NO \longrightarrow 2NO_2 \tag{32}$$

Recent nighttime measurements of NO_3, by a differential optical absorption technique,[41-42] have found levels of up to 200 ppt NO_3 present in the polluted urban boundary layer. This has initiated a series of laboratory studies of NO_3 reactions with aldehyde, alkane, alkene, and aromatic compounds.[43-44] These reactions are observed to be considerably faster than the corresponding ozone reactions. The reactions of NO_3 with aldehydes and alkanes are simple abstraction reactions, emphasizing the radical character of NO_3

$$NO_3 + R(C=O)H \longrightarrow R(C=O) + HONO_2 \tag{33}$$

$$NO_3 + RH \longrightarrow R + HONO_2 \tag{34}$$

The mechanism of the reactions with alkenes and aromatics are not well established. It is speculated that the reactions of NO_3 with hydrocarbons introduce a new and previously neglected radical chemistry in the nighttime atmosphere, yielding PAN and other typical photooxidation products, in the absence of sunlight. It must, however, be recognized that the high nighttime concentrations of NO_3 observed in the Los Angeles atmosphere may not be typical of most urban polluted atmospheres. Recent field measurements in Germany often failed to detect NO_3 during the night, suggesting that rapid hydrolysis of N_2O_5 in fog droplets and on wet aerosols can be a significant sink for NO_3. Concentrations of this important reservoir species cannot be measured in situ by differential optical absorption, but must be inferred from (NO_2), (NO_3) and the equilibrium constant.[45] While the homogeneous reaction of N_2O_5 with water vapor is extremely slow, hydrolysis in the water film of wet aerosol particles and/or fog droplets seems to be efficient, becoming increasingly important as a nighttime loss mechanism for NO_x during the winter season, when the loss rate via the reaction of HO with NO_2 is low due to low levels of HO as a result of reduced photochemical activity in the atmosphere. Clearly further work is needed to quantify the nighttime chemistry of NO_3 and N_2O_5.

Recent developments in instrumentation technology for measuring nitrogen total reactive species (NO_y) in the troposphere[46-48] indicate that a substantial portion (up to 40%) of the reactive odd-nitrogen cannot be accounted for from the speciated nitrogen budget (NO, NO_2, HNO_3, PAN, and particulate nitrate). Since the photochemical oxidation cycle is intimately coupled to our understanding of the nitrogen chemistry, it is extremely important that we determine these unknown species and their mechanistic pathways.

The above sequence of steps, although simplified, contains the essential features of the NO to NO_2 oxidation process and the subsequent formation of ozone in the ambient atmosphere.

The rate and yield of oxidant formation depends on many factors. Some important examples include: the chemical reactivity of the hydrocarbons present; solar intensity; temperature and the presence of free radical initiating compounds. In the ambient atmosphere, the photolysis of aldehydes, nitrous acid, and O_3 are important initial sources of radicals and play a significant role in the net accumulation of ozone in the polluted atmosphere. During the course of the overall photooxidation process, the free radical pool is maintained by several sources as has been demonstrated in the reaction sequences discussed above. Since the reaction of free radicals with NO form a cyclic process, any additional source of radicals will add to the pool and increase the cycle rate. Conversely, any reaction that removes free radicals will slow the cycle rate. This cyclic process of photooxidation in the polluted atmosphere is presented in schematic form in Figure 4.[37]

In its simplest form the photochemical oxidation cycle in the polluted atmosphere is governed by the following basic features. Hydrocarbon/organic pollutants in the atmosphere are attacked by free radical species initiated by a select group of compounds which are, for the most part, activated by sunlight. After the initial free radical attack, the hydrocarbon/organic compounds decompose through paths resulting in the production of peroxy radical species (HO_2, RO_2) and partially oxidized products which in themselves can be photoactive radical producing compounds. The peroxy radicals react with NO, producing NO_2, and in the process also forming hydroxy and organic oxy radical species (HO, RO). The RO radicals can further oxidize, forming additional peroxy radicals and more partially oxidized products, thereby completing the inner cycle of the reaction chain process illustrated in Figure 4. Alternatively, the HO radical can attack the hydrocarbon/organic pool present in the polluted atmosphere, completing the outer cycle of the reaction chain process. The resultant effect in either case is the conversion of NO to NO_2 and the oxidizing of hydrocarbon/organic compounds present in the atmosphere. The complex mixture of hydrocarbon/organic compounds present in the polluted atmosphere react at different rates depending upon their molecular structure, the result being varying yields of free radical

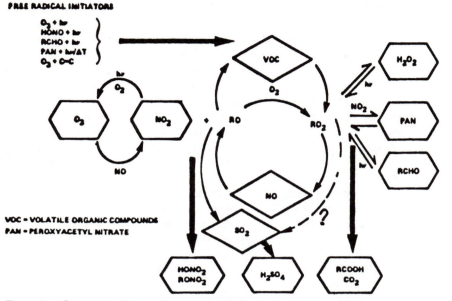

Figure 4. Schematic of the polluted atmospheric photooxidation cycle.

species, ozone, NO_2, PAN, and other partially oxidized organic products as a function of the composition of the hydrocarbon/organic compounds present, their levels and ratio with NO_x.

The organic compounds important in the chemistry of the troposphere include: alkanes, alkenes, aromatics, and terpenes, as well as oxygenated species such as aldehydes, ketones, esters, ethers, and alcohols. There are a very large number of chemical reactions that can take place among these organic compounds and the free radical species discussed above. The details of these processes have been explored through computer simulation.[5-11] Reaction pathways for the decomposition of hydrocarbons via hydroxy radical, a major oxidation route, have been studied extensively.[5-6,9,28],[49-52] Uncertainties still remain regarding the mechanistic pathways for many classes of compounds. For example, in the case of aromatic compounds, a large fraction of the carbon budget, assessed through product analysis, remains unaccounted for, thus leaving the oxidation potential and secondary product formation of these compounds very uncertain.

APPLICATION OF CHEMICAL MECHANISMS

The Environmental Protection Agency's interest in the research and development of chemical mechanisms of polluted atmospheres stems from their responsibility to manage air quality and its associated environmental effects. The first oxidant control strategies recommended by the U.S. Environmental Protection Agency were based on ambient pollutant concentration data taken in several U.S. cities. These data were analyzed to relate the daily one-hour averaged maximum oxidant concentration to the average nonmethane hydrocarbon (NMHC) concentration observed in the 6 to 9 a.m. time period.[53-54] This so-called upper-limit curve, which inherently assumes that hydrocarbons are the unique dependent variable in oxidant control, became the basis for calculating the control requirement necessary to meet the National Ambient Air Quality Standard for oxidants. The resulting oxidant-NMHC relationship, referred to as Appendix J, was determined to be scientifically incomplete, based on evolving knowledge of atmospheric photochemical oxidation processes[55] and the meteorology of urban boundary layers.[56]

With improved knowledge has come the development of a variety of improved scientific approaches for quantitatively treating the relationship between precursor emissions and ozone formation. Among others, these include, in order of increasing meteorological sophistication: (1) the Empirical Kinetics Modeling Approach;[57] (2) the Photochemical Box Model;[58-59] (3) the Lagrangian Photochemical Model;[60] and (4) the Urban Airshed Model.[61] All these advanced techniques incorporate state of the art photochemical mechanisms.

The chemical mechanisms are a critical component in modeling techniques which provide a quantitative relationship between the emission of chemical precursors which react in the atmosphere both in the gas- and liquid-phases and in sunlight as well as in the dark, to produce chemical species of environmental concern. Currently these compounds include: ozone, nitrogen dioxide, fine particulate matter (primarily sulfates), and acid-bearing substances, and most likely the list will expand with time. Relevant to this discussion is the modeling of the formation of ozone in urban atmospheres. The air quality simulation model incorporates a chemical mechanism in conjunction with emissions information and some treatment of the transport and diffusion of the chemical species under study. The model is exercised to provide quantitative guidance as to the amount of precursor emission control (nonmethane hydrocarbons and oxides of nitrogen) that is required to meet a specified concentration of ozone. At issue is the precision and accuracy of this quantitative relationship, methods for its evaluation, and standards of acceptability (or success).

In recognition of these needs the scientific community has, on a continuing basis, been developing and evaluating mechanisms against smog chamber data sets. As the science of chemical mechanism development has become more sophisticated, the community's requirements for quality and performance have also become more refined. The feedbacks and intercomparisons that are components of the methodological development process, began to identify limitations in the chamber experiments. Scientists began to question the degree of effort one should make to fit individual runs, series of runs, and runs between different experimental smog chamber systems. The debate arose from the fact that many smog chambers whose data sets had been used for mechanism development had not been adequately characterized with respect to wall effects. The chamber walls which can act as both sources and sinks for important chemical constituents introduce noise in the chamber data.[62-63] This limits the precision and accuracy claims which might be inferred by precise fits of modeled and observed concentration-time profiles.

In addition, these limitations are not always explicitly characterizable, and are thought to ultimately contaminate mechanisms which have been developed from them. This contamination can take the form of inherent noise or a systematic bias. This phenomenon represents a significant scientific challenge to the community and has important implications on the limits of precision and accuracy of the quantitative relationship desired in regulatory applications. If the chamber effects are unknown or incorrectly specified, the result is a chemical mechanism that has most likely over- or under-compensated the radical production processes in order to achieve an acceptable fit of the chamber data. The application of such a mechanism in a regulatory model will result in a systematic bias in the quantitative relationship between precursor emissions and ozone production.

The utilization of atmospheric observations as a vehicle for mechanism development and evaluation is intuitively the most satisfying scientifically. However, until recently, the instrumentation technology necessary to characterize the detailed chemical components of the atmospheric system were beyond reach. Also the complexities introduced by the dynamics of the atmosphere introduced considerable uncertainties which make diagnostic interpretations of mechanisms quite difficult. Yet even with these caveats it would seem that progress in instrumentation technology and the importance of studying real world chemical systems suggests that atmospheric observations of increasing sophistication become a major component in the development of new generation chemical mechanisms.

SCIENTIFIC VERSUS REGULATORY SUCCESS

It is quite apparent that the precision and accuracy requirements which would meet the scientific community's standards for success may not be acceptable to the regulatory community. By this I mean that the regulatory application, in order for it to be an effective tool for developing quantitative control strategies, may require better precision and accuracy performance from the mechanism than the science can provide.

For example, the scientific community in reviewing the various sources of error associated with the chemical mechanism development process anticipates uncertainties of the order of plus or minus 30% in the mechanism's predictive capability when compared with chamber data.[64-66] Establishing that this is a reasonable error limit when the mechanism is applied under real atmospheric conditions remains a critical issue which must be demonstrated if these approaches are to have any scientific credibility. More importantly, if the 30% uncertainty is reflected in the control requirement of precursors (e.g., nonmethane hydrocarbons) to meet the ozone standard in a given city, the associated cost differentials can be both economically and socially prohibitive.

The recognition of this uncertainty and factoring it into the quantitative procedures required in state implementation plans to demonstrate a course of action for attaining the national ambient air quality standard for ozone would seem a logical first step to be taken by the regulatory community. The next step should be to establish realistic precision and accuracy performance standards for mechanisms to be used in quantitative models for control strategy development.

TRANSPORTED OZONE AND OZONE PRECURSORS

The chemical and physical characteristics of pollutant species to a large extent govern the relative importance of the transport, transformation, and removal processes that they undergo. For example, in the case of modeling ozone episodes in urban environments, the time scale of interest is of the order of a diurnal cycle (≈ 12 hrs). During periods of high solar insolation and temperature and limited atmospheric ventilation of the urban area, the production of ozone is rapid. Because removal of pollutant gases by dry deposition is a relatively slow process, its impact on modeling ozone on the urban scale is minor. In the case of wet removal, the process is not of interest, since precipitating days lack the solar intensity to drive the photochemical smog formation process. Neither of these removal processes can be considered unimportant when considering oxidant and its precursors on the regional scale; that is, transport scales of the order of 1000 km and chemical and physical process time scales of the order of several days. It should also be noted that regional contributions (lateral and top boundary inflow) of ozone to urban episodes are significant under many atmospheric conditions, and must be considered in the analyses performed on urban areas.

The potential for significant contributions to ozone production and that of other photochemical oxidants on a hemispheric scale due to nonmethane hydrocarbon and NO_x emissions resulting from man's activities is quite real. The state of the science is such that a quantitative understanding of this contribution is not yet possible. Some phenomenological processes speculated to be important in large-scale ozone/oxidants contributions are: (1) transported ozone/oxidants formed within the confines of the urban plume and dispersed over regional and continental scales; (2) the formation of ozone/oxidants involving photochemical reactions of the remnants of aged urban air masses (partially oxidized hydrocarbons and slower reacting hydrocarbon species) which have been transported and dispersed over multiple days; and (3) the formation of ozone/oxidants from photochemical reactions of low-level emissions of nonmethane hydrocarbons and NO_x over the continent due to man as well as nature.

Our limited understanding of these ozone/oxidants production sources is somewhat traceable to difficulties in studying the reaction pathways and ozone formation potentials of the slower oxidizing hydrocarbon species and the characterization of nonmethane species composition and NO$_x$ levels in aged polluted air masses and background continental air.

CONCLUSIONS

Progress continues to be made in the physical and chemical characterization of the complex processes associated with the production of oxidants/O$_3$ in polluted atmospheres. Our goal is to provide tools based on sound, credible scientific principles which relate primary precursor emissions to oxidants/O$_3$ products. As a result of continuing research and development efforts, significant advances in our understanding of the complex processes operating in photochemical smog have been made. These include improved mechanisms for treating the photochemical oxidant cycle, better understanding and treatment of urban boundary layer processes, initiation of systematic model performance evaluation studies, advances in sensitivity and uncertainty analysis techniques and their application, and consideration of oxidants/O$_3$ on regional scales (≈ 1000 km). In addition, the process of determining the nature and extent of the uncertainty associated with model prediction has begun. The recognition of model uncertainty and its estimation requires that the scientific community make a special effort to better articulate the limitations in model predictability and to provide a better understanding of how these limits can be incorporated effectively in the decision making process.

In order to improve the precision and accuracy of chemical mechanisms, and therefore their ability to provide quantitative relationships between precursor emissions and ozone production, it is recommended that:

- procedures be developed for mapping chemical mechanisms in order to provide flow diagrams of all mechanisms under research and development or currently used in applications to identify critical nodes for radical initiation, propagation, and termination;
- laboratory studies be executed to reconcile differences in critical nodes between mechanisms which result from lack of knowledge of the chemical details of a reaction process (e.g., elementary reaction steps, fragmentation channels and yields);
- guidelines and standards be developed for smog chamber operations and basic requirements for chamber characterization and requirements that chamber performance characteristics must be documented and submitted to peer review, preferable through publication in the open literature.
- guidelines be developed and smog chamber intercomparison studies performed to determine internal bias among systems, variations in artifact effects and overall reproducibility of results between systems.
- programs be initiated to systematically build an atmospheric chemical observational database for the purpose of diagnostic interpretation and evaluation of mechanisms.

REFERENCES

1. Logan, J. A. "Tropospheric Ozone: Seasonal Behavior, Trends, and Anthropogenic Influence," *J. Geophys. Res.*, 90(10): 463-510 (1985).
2. Leighton, P. A. *Photochemistry of Air Pollution* (New York: Academic Press, Inc., 1961).
3. Seinfeld, J. H. *Atmospheric Chemistry and Physics of Air Pollution*, (New York: John Wiley & Sons, Inc., 1985).
4. Finlayson-Pitts, B. J., and J. N. Pitts, Jr. *Atmospheric Chemistry: Fundamentals and Experimental Techniques* (New York: John Wiley & Sons, Inc., 1986).
5. Demerjian, K. L., J. A. Kerr, and J. G. Calvert. "The Mechanisms of Photochemical Smog Formation," *Adv. Environ. Sci. Technol.*, 4:1-262 (1974).
6. Carter, W. P. L., A. C. Lloyd, J. L. Sprung, and J. N. Pitts, Jr. "Progress in the Validation of a Detailed Mechanism for the Photooxidation of Propene and n-Butane in Photochemicals," *Int. J. Chem. Kinet.*, 11:45-111 (1979).
7. Baldwin, A. C., R. Baker, D. M. Golden, and D. G. Hendry. "Photochemical Smog. Rate Parameter Estimates and Computer Simulation," *J. Phys. Chem.*, 81:2483 (1977).
8. Atkinson, R., A. C. Lloyd, and L. Winges. "An Updated Chemical Mechanism for Hydrocarbon/NO_x/SO_2 9. Falls, Photo-Oxidations Suitable for Inclusion in Atmospheric Simulation Models," *Atmos. Environ.* 16:1341-1355 (1982).
9. Falls, A. H., and J. H. Seinfeld. "Continued Development of a Kinetic Mechanism for Photochemical Smog," *Environ. Sci. Technol.*, 12:1398 (1978).
10. Whitten, G. Z., H. Hogo, and J. P. Killus. "The Carbon Bond Mechanism: A Condensed Kinetic Mechanism for Photochemical Smog," *Environ. Sci. Technol.* 14:690-700 (1980).
11. Niki, H., E. E. Daby, and B. Weinstock. "Mechanism of Smog Reactions," *Adv. Chem. Ser.* 113:16 (1972).
12. Benson, S. W. *Thermochemical Kinetics* (New York: John Wiley & Sons, 1968).
13. Pitts, J. N., Jr., A. C. Lloyd, and J. L. Sprung. "Ecology, Energy and Economies," *Chem. Br.* 11:247 (1975).
14. Baulch, D. L., R. A. Cox, R. F. Hampson, Jr., J. A. Kerr, J. Troe, and R. T. Watson. "Evaluated Kinetic and Photochemical Data for Atmospheric Chemistry," *J. Phys. Chem. Ref. Data* 9:295 (1980).
15. Demerjian, K. L., K. L. Schere, and J. T. Peterson. "Theoretical Estimates of Active (Spherically Integrated) Flux and Photolytic Rate Constants of Atmospheric Species in the Lower Troposphere," *Adv. Environ. Sci. Technol.*, 10:369-459 (1980).
16. Baulch, D. L., R. A. Cox, P. J. Crutzen, R. F. Hampson, Jr., J. A. Kerr, J. Troe, and R. T. Watson. "Evaluated Kinetic and Photochemical Data for Atmospheric Chemistry: Supplement I. CODATA Task Group on Chemical Kinetics," *J. Phys. Chem. Ref. Data* 11:327-496 (1982).
17. Stedman, D. H., and J. O. Jackson. "The Photostationary State in Photochemical Smog," *Int. J. Chem. Kinetics* 1:493-501 (1975).
18. Calvert, J. G. "Test of the Theory of Ozone Generation in Los Angeles Atmosphere," *Environ. Sci. Technol.* 10:248-256 (1976).
19. Peterson, J. T., and K. L. Demerjian. "The Sensitivity of Computed Ozone Concentrations to U.V. Radiation in the Los Angeles Area," *Atmos. Environ.* 10:459-468 (1976).

20. Chan, W. H., R. J. Nordstrom, J. G. Calvert and J. H. Shaw. "Kinetic Study of HONO Formation and Decay Reactions in Gaseous Mixtures of HONO, NO, NO_2, H_2O and N_2," *Environ. Sci. Technol.* 10:674–682 (1976).

21. Harris, G. W., W. P. L. Carter, A. M. Winer, J. N. Pitts, Jr., U. Platt, and D. Perner. "Observations of Nitrous Acid in the Los Angeles Atmosphere and Implications for Predictions of Ozone-Precursor Relationships," *Environ. Sci. Technol.* 16:414–419 (1982).

22. Sjodin, A. and M. Ferm. "Measurements of Nitrous Acid in an Urban Area," *Atmos. Environ.* 19:985–992 (1985).

23. Duce, R. A., V. A. Mohnen, P. R. Zimmerman, D. Grosjean, W. Cautreels, R. Chatfield, R. Jaenicke, J. A. Ogren, E. D. Pellizzari, and G. T. Wallace. "Organic Material in the Global Troposphere," *Reviews of Geophysics and Space Physics,* 21: 921–952, (1983).

24. Singh, H. B., and L. J. Salas. "Measurement of Selected Light Hydrocarbons Over the Pacific Ocean: Latitudinal and Seasonal Variations," *Geophys. Res. Letters* 9:842–845 (1982).

25. Ehhalt, D. H., J. Rudolph, F. Meixner, and U. Schmidt. "Measurements of Selected C_2-C_5 Hydrocarbons in the Background Troposphere: Vertical and Latitudinal Variations," *J. Atmos. Chem.* 3:29–52 (1985).

26. Singh, H. B., L. J. Salas, B. K. Cantrell, and R. M. Redmond. Distribution of Aromatic Hydrocarbons in the Ambient Air," *Atmos. Environ.* 19:1911–1919 (1985).

27. Roberts, J. M., R. S. Hutte, F. C. Fehsenfeld, D. L. Albritton, and R. E. Sievers. "Measurements of Anthropogenic Hydrocarbon Concentration Ratios in the Rural Troposphere: Discrimination Between Background and Urban Sources," *Atmos. Environ.* 19:1945–1950 (1985).

28. Calvert, J. G., and S. Madronich. "Theoretical Study of the Initial Products of the Atmospheric Oxidation of Hydrocarbons," *J. Geophys. Res.,* 92:2211–2220 (1987).

29. Graham, R. A., A. M. Winer, and J. N. Pitts, Jr. "Temperature Dependence of the Unimolecular Decomposition of Pernitric Acid and Its Atmospheric Implications," *Chem. Phys. Letters,* 51:215–220 (1977).

30. Darnell, K. R., W. P. L. Carter, A. M. Winer, A. C. Lloyd, and J. N. Pitts, Jr. "Importance of RO_2 + NO in Alkyl Nitrate Formation from C_4-C_6 Alkane Photooxidations Under Simulated Atmospheric Conditions," *J. Phys. Chem.* 80:1984–1950 (1976).

31. Lonneman, W. A., J. J. Bufalini, and R. L. Seila. PAN and Oxidant Measurements in Ambient Atmospheres," *Environ. Sci. Technol.* 10:374–380 (1976).

32. Singh, H. B., and L. J. Salas. "Peroxy Peroxyacetyl Nitrate in the Free Troposphere," *Nature* 302:326–328 (1983).

33. Spicer, C. W., M. W. Holdren, and C. W. Keigley. "The Ubiquity of Peroxyacetyl Nitrate in the Continental Boundary Layer," *Atmos. Environ.* 17:1055–1058 (1983).

34. Brice, K. A., S. A. Penkett, D. H. F. Atkins, F. J. Sandalls, D. J. Bamber, A. F. Tuck, and G. Vaughan. "Atmospheric Measurements of Peroxyacetylnitrate (PAN) in Rural, South-East England: Seasonal Variations, Winter Photochemistry, and Long-Range Transport," *Atmos. Environ.* 18:2691–2702 (1984).

35. Singh, H. B., and P. L. Hanst. "Peroxyacetyl Nitrate (PAN) in the Unpolluted Atmosphere: An Important Reservoir for Nitrogen Oxides," *Geophys. Res. Lett.* 8:941–944 (1981).

36. Aikin, A. C., J. R. Herman, E. J. R. Maier, and C. J. McQuillan. "Influence of Peroxyacetyl Nitrate (PAN) on Odd Nitrogen in the Troposphere and Lower Stratosphere," *J. Geophys. Res.* 87:3105–3118 (1982).
37. Demerjian, K. L. "The Atmospheric Photooxidation Cycle and the Influence of Tropospheric Pollution on Ozone," *Proceedings of the Quadrennial International Ozone Symposium,* Vol. I, August 4–9, 1980, Boulder, CO.
38. Morris, E. D., and H. Niki. "Reaction of Dinitrogen Pentoxide with Water," *J. Phys. Chem.* 77:1929–1932 (1973).
39. Tuazon, E. C., R. Atkinson, C. N. Plum, A. M. Winer, and J. N. Pitts, Jr. "The Reaction of Gas Phase N_2O_5 with Water Vapor," *Geophys. Res. Lett.* 10:953–956 (1983).
40. Russell, A. G., G. J. McRae, and G. R. Cass. "The Dynamics of Nitric Acid Production and the Fate of Nitrogen Oxides," *Atmos. Environ.* 19:893–903 (1985).
41. Platt, U. F., Dieter Perner, A. M. Winer, G. W. Harris, J. N. Pitts, Jr. "Detection of NO_3 in the Polluted Troposphere by Differential Optical Absorption," *Geophys. Res. Lett.* 7:89–92 (1980).
42. Platt, V. F., A. M. Winer, H. W. Biermann, R. Atkinson, and J. N. Pitts, Jr. "Measurements of Nitrate Radical Concentrations in Continental Air," *Environ. Sci. Technol.,* 18:365–369 (1984).
43. Atkinson, R., S. M. Aschmann, A. M. Winer, and J. N. Pitts, Jr. "Kinetics of the Gas-Phase Reactions of NO_3 Radicals with a Series of Dialkenes, Cycloalkenes, and Monoterpenes at 295 + 1K," *Environ. Sci. Technol.* 18:370–375 (1984).
44. Atkinson, R., C. N. Plum, W. P. L. Carter, A. M. Winer, and J. N. Pitts, Jr. "Rate Constants for the Gas-Phase Reactions of NO_3 Radicals with a Series of Organics in Air at 298 $\pm$ 1K, *J. Phys. Chem.* 88:1210–1215 (1984).
45. Malko, M. W., and J. Troe. "Analysis of the Unimolecular Reaction N_2O_5 + M —> NO_2 + NO_3 + M," *Int. J. Chem. Kinet.,* 14: 399 (1982).
46. Fahey, D. W., C. S. Eubank, G. Hubler, and F. C. Fehsenfeld. "Evaluation of a Catalytic Reduction Technique for the Measurement of Total Reactive Odd-Nitrogen NO_x in the Atmosphere," *J. Atmos. Chem.* 3:435–468 (1985).
47. Dickerson, R. R. "Measurement of Reactive Nitrogen Compounds in the Free Troposphere," *Atmos. Environ.,* 18:2585–2593 (1984).
48. Fahey, D. W., G. Huber, D. D. Parrish, E. J. Williams, R. B. Norton, B. A. Ridley, H. B. Singh, S. C. Liu, and F. C. Fehsenfeld. *J. Geophys. Res.* 91:9781–9793 (1986).
49. Perry, R. A., R. Atkinson, and J. N. Pitts, Jr. "Kinetics and Mechanism of the Gas Phase Reactions of OH Radicals with Aromatic Hydrocarbons Over the Temperature Range 296–435 K," *J. Phys. Chem.,* 82:296–304 (1977).
50. Atkinson, R., and A. C. Lloyd. *J. Phys. Chem. Ref. Data,* 13:315–444 (1984).
51. Tuazon, E. C., R. Atkinson, H. MacLeod, H. W. Biermann, A. M. Winer, W. P. L. Carter, and J. N. Pitts, Jr. "Yields of Glyoxal and Methylglyoxal from the NO_x-Air Photooxidations of Toluene and m- and p- Xylene," *Environ. Sci. Technol.* 18:981–984 (1984).
52. Grosjean, D. "Atmospheric Reactions of Ortho Cresol: Gas Phase and Aerosol Products," *Atmos. Environ.,* 18:1641–1652 (1984).
53. U.S. Environmental Protection Agency, "Air Quality Criteria for Nitrogen Oxides", Air Pollution Control Office Publication No. AP-84, Washington, D.C. (January 1971).

54. *Federal Register,* 36:115486 (August 14, 1971).
55. Dimitriades, B. "Oxidant Control Strategies. Part I. Urban Oxidant Control Strategy Derived from Existing Smog Chamber Data," *Environ. Sci. Technol.* 11:80 (1977).
56. Demerjian, K. L. "Photochemical Diffusion Models for Air Quality Simulation: Current Status," in *Assessing Transportation-Related Air Quality Impacts,* Special Report 167, Transportation Research Board, National Research Council, Washington, DC (1976), pp. 459–469.
57. Meyers, E. L., J. E. Summerhays, and W. P. Freas. "Uses, Limitations and Technical Basis of Procedures for Quantifying Relationships Between Photochemical Oxidants and Precursors," EPA-450/2-77- 021a, U.S. Environmental Protection Agency, Research Triangle Park, NC (1977).
58. Demerjian, K. L., and K. L. Schere. "Applications of a Photochemical Box Model for O₃ Air Quality in Houston, Texas," *Proceedings, Ozone/Oxidants: Interaction with the Total Environment II. Houston, TX,* 14–17 *October* 1979, APCA, Pittsburgh, PA (1979), p. 329.
59. Schere, K. L., and K. L. Demerjian. "User's Guide for the Photochemical Box Model (PBM)," EPA-600/8-84-022a, U.S. Environmental Protection Agency, Research Triangle Park, NC (1984).
60. Lurmann, F., D. Godden, A. C. Lloyd, and R. A. Nordsieck. "A Lagrangian Air Quality Simulation Model-Vol I. Model Formulation; Vol II. User's Manual," EPA-600/8-79-015a and 015b, U.S. Environmental Protection Agency, Research Triangle Park, NC (1979).
61. Reynolds, S. D., P. M. Roth, and J. H. Seinfeld. "Mathematical Modeling of Photochemical Air Pollution—I. Formulation of the Model," *Atmospheric Environment* 7:1033 (1973).
62. Carter, W. P. L., R. Atkinson, A. M. Winer, and J. N. Pitts, Jr. "Evidence for Chamber-Dependent Radical Sources: Impact on Kinetic Computer Models for Air Pollution," *Int. J. Chem. Kinet.,* 13:735 (1981).
63. Carter, W. P. L., R. Atkinson, A. M. Winer, and J. N. Pitts,. Jr. "Experimental Investigation of Chamber-Dependent Radical Sources," *Int. J. Chem. Kinet.,* 14:1071 (1982).
64. Falls, A. H., G. J. McRae, and J. H. Seinfeld. "Sensitivity and Uncertainty of Reaction Mechanisms for Photochemical Air Pollution," *Intern. J. Chem. Kinet.,* 1137 (1979).
65. Dunker, A. M., S. Kumar, and P. H. Berzins. "A Comparison of Chemical Mechanisms Used in Atmospheric Models," *Atmospheric Environment* 18:311 (1984).
66. Shafer, T. B., and J. H. Seinfeld. "Comparative Analysis of Chemical Reaction Mechanisms for Photochemical Smog—II. Sensitivity of EKMA to Chemical Mechanism and Input Parameters," *Atmos. Environ.,* 20:487–499 (1986).

CHAPTER 2

Transport of Ozone and Its Implications for New England

Chi Ho Sham

INTRODUCTION

In recent years, ozone in the planetary boundary layer (i.e., the lowermost part of the atmosphere) has become a major environmental issue. While levels of most major air pollutants have been decreasing during the past decade, surface ozone concentrations across the heavily populated part of the United States have not shown any significant decline. Moreover, the National Ambient Air Quality Standard (NAAQS) established by the U.S. Environmental Protection Agency for ozone (i.e., 120 ppb) has been exceeded on a regular basis. Ozone is neither generated in high concentrations in the lower atmosphere by natural processes nor emitted directly by the combustion of fossil fuels and other chemical processes. It is produced as a secondary pollutant from photochemical reactions among hydrocarbons and nitrogen oxides in the presence of sunlight. Since a direct control strategy to reduce ozone is not available, numerous studies have been conducted in an attempt to reveal the underlying processes and factors that produce elevated ozone concentrations at both urban and nonurban locations in the United States. Hopefully, an effective scheme can be developed in the near future to control the formation of ozone in the boundary layer.

The possible sources of surface ozone at a given locality are: (1) transport of stratospheric ozone through the tropopause into the upper troposphere and subsequently into the boundary layer; (2) in situ photochemical ozone formation; (3) transport of ozone and its precursor within pollution plumes from heavily polluted areas and industrial areas on a subregional scale; and (4) transport of ozone and its precursors within polluted air masses on a regional scale. However, due to the heterogeneous nature of these sources in both space and time, the degree of contribution by these above sources to surface ozone concentrations at a particular locality is highly uncertain and very difficult to quantify.

The overall contribution by the transport of stratospheric ozone into the boundary layer is believed to be very small. Estimated values of mean stratospheric ozone contribution to surface ozone concentration range from 7 to 12 ppb.[1,2] Since the control of stratospheric ozone flux is currently unattainable based on available technologies, the reduction of ozone can only be achieved by dealing with other ozone sources.

Factors affecting in situ formation of ozone have been dealt with elsewhere in this book, and thus will not be repeated here. However, relevant information that is pertinent to the understanding of ozone transport will be discussed. First, the typical daily pattern of surface ozone concentrations at a midlatitude location is cyclical with a nocturnal minimum and an afternoon maximum. This diurnal cycle is controlled by the availability of solar radiation for photochemical reactions. Second, it can also be noted that urban sites are different from nonurban sites in their ozone behavioral patterns. It is a generally agreed-upon fact that urban sites are associated with lower ozone peak values due to an abundance of scavenging emissions such as nitrogen oxides and hydrocarbons. Third, apart from the diurnal variation of ozone concentration, a noctural radiation inversion is commonly observed for inland locations in the midlatitudes.[3] Such a radiation inversion is the key to the establishment of two ozone regimes, one at the ground level and one aloft over the radiation inversion. Ozone concentrations tend to be higher above the radiation inversion because of the lack of scavenging compounds. However, during early morning, surface warming by the sun dissipates the radiation inversion and thus permits the mixing of the air from aloft with surface air. Such a process can cause an increase of ozone concentration at the ground level during the morning and afternoon hours.

In this chapter, ozone transport within urban plumes and other pollution plumes, along with ozone transport within air masses of high pressure systems are briefly reviewed. In addition, the implications of ozone transport for New England are discussed.

OZONE TRANSPORT IN NEW ENGLAND

Ozone Concentrations in Urban Plumes and Other Plumes

Ozone transport within urban plumes has been studied extensively using aircraft tracking, ground level mapping, and mathematical modeling. Ozone measurements on board aircraft have been made on plumes from a number of cities, including New York City and Boston. A list of selected examples of these measurements is presented in Table 1.

By examining Table 1, it can be seen that the transport of ozone and its precursors within urban plumes can significantly alter the ozone budgets of areas downwind of the source regions. During episodes of high ozone concentrations, sizeable areas downwind of major pollution sources will experience elevated ozone levels for days. Spicer et al.[4] examined the urban plume from New York City and concluded that transport of ozone is most effective over ocean due to the lack of scavenging emissions and smooth terrain. It was observed that the New York plume can extend over 240 to 400 km downwind over the ocean. In addition, the vertical dimension of these urban plumes can be quite substantial (e.g., up to 600 meters in altitude). The transport of ozone over land is much more variable and is strongly controlled by wind conditions. In the study by Spicer et al.,[4] the dimensions of urban plumes were found to vary from 30 to 130 km in width and 160 to 280 km in length. Spicer et al.[4] further pointed out that ozone and its precursors often survive from one day to the next by staying aloft above the nocturnal radiation inversion. By staying aloft, fossil ozone and other ozone precursors are isolated from scavenging by ground surface and surface-based scavenging emissions. This ozone and other ozone precursors aloft subsequently mix downward with the breakup of the nocturnal inversion during the second day. The occurrences of high levels of ozone after nightfall along with the occurrences of high ozone concentrations during early morning hours in nonurban areas all point to the influence of ozone transport on ambient ozone concentrations downwind of major ozone sources.

Table 1.   Ozone Concentrations within Urban Plumes of New York City, New York and Boston, Massachusetts.

Source Area	Date	Ozone concentration (ppb)		Distance Downwind	Reference
		Within Plume	Outside of Plume		
New York City	8/9/75	200	60	100 km	No. 4
New York City	7/23/75	250–300	70–85	125 km	No. 4
		150–200	80–90	200 km	No. 4
		100–150	60–70	300 km	No. 4
New York City	7/24/75	130	na	350 km	No. 4
		145	na	450 km	No. 4
New York City	8/10/75	130	na	160 km	No. 4
New York City	8/6/80	352	<120	125 km	No. 5
Boston	8/6/80	231	<120	55 km	No. 5

Note: na = not available.

In a more recent study by Spicer et al.,[5] dimensions and ozone concentrations of the New York City plume and the Boston plume were again measured using aircraft tracking. The width for the New York plume was estimated to be about 60 to 65 km, whereas the estimated width of the Boston plume was about 35 to 45 km. The maximum ozone concentration in the New York City plume was 352 ppb and was measured at approximately 125 km downwind from New York City over Long Island Sound on August 6, 1980. As for the Boston plume, the maximum ozone concentration was 231 ppb and was observed 55 km downwind near Newburyport, Massachusetts on August 5, 1980. Clearly, urban plumes of very high ozone concentrations can be detected downwind of urban areas such as New York City and Boston. Ozone concentrations within the studied plumes can exceed 120 ppb up to several hundred kilometers downwind of the urban areas.

Apart from aircraft tracking studies, a study by Rubino et al.[6] has demonstrated that surface measurements of ozone concentrations can be used to reveal the transport of ozone within urban plumes. Rough plots of ozone isopleths constructed from measurements over rural areas of Connecticut have been used to trace the June 10, 1974 ozone episode across the state. The elevated ozone concentrations were associated with an air parcel that was over New York City during morning peak traffic. To a certain extent, this particular approach may yield valuable information regarding the movement of urban plumes at the ground level. Obviously, such an approach is affected by the coarseness of the monitoring network and is not capable of capturing the effects of nocturnal inversion and surface scavenging.

A number of Lagrangian trajectory models have also been applied to study the formation and destruction of ozone in urban plumes. By incorporating (a) the Hydrocarbons (HC):Nitrogen Oxides (NO$_x$) ratios, (b) plume emergence times and (c) meteorological conditions, Bazzell and Peters[7] predicted that for a plume emerging at 8:30 a.m. into an atmosphere with a 12 km/hr wind, a fourfold decrease in the hydrocarbons concentration is needed to produce a decrease of ozone concentration from 240 ppb to 30 ppb at approximately 80–90 km downwind. The model also predicted that with the highest HC:NO$_x$ ratio, the ozone concentration at 114 km downwind (at 6:00 p.m.) would be 200 ppb. In the model developed by Spicer and Sverdrup,[8] meteorological parameters were introduced to present varying dilution rates for both daylight and nighttime periods. Based on their results, the authors concluded that much NO$_x$ could cycle through peroxyacyl nitrates (PANs) to form NO$_x$ in the following day.

In addition to studies of urban plumes from a single major urban area, other studies have indicated that the overlapping of multiple plumes (e.g., from two or more cities) may be of great importance in affecting ozone concentrations downwind. For example, Sexton[9] has examined the additive effect of ozone plumes from a number of small cities in Wisconsin and Illinois (populations ranging from 35,000 to 173,000) and has found that pollutant emissions from these small cities are substantial with respect to downwind ambient ozone concentrations. The combined plume from these small cities was typically 10–50 km in width with an ozone concentration 20 to 30 ppb above background level.

In a similar type of study for the combined Baltimore-Washington, DC plume, Westberg[10] concluded that elevated ozone concentrations up to 120 to 220 ppb were found in areas with maximum plume overlap. In addition, elevated ozone levels were observed in areas as far as 190 km downwind from Baltimore.

The two cases of interurban transport presented by Spicer et al.[4] and Spicer et al.[5] discussed above merit further attention. During the July 23–24, 1975 and July 31–August 1, 1980 ozone episodes, ozone plumes from New York City were tracked across southern New England. In both instances, the plumes reached Boston within one day. During the second day, the additive effect of the two urban plumes generated elevated ozone concentrations both locally in the Boston vicinities and downwind of Boston. For example, on July 24, 1975, high concentrations of ozone (up to 145 ppb) were detected northeastward of Boston over the Atlantic Ocean and at coastal locations northeast of Portland, Maine. As for the July 31–August 1, 1980 episode, the background levels of ozone within the Boston area boundary layer increased from 50 to 60 ppb on July 31 to over 100 ppb on August 1.

Elevated ozone concentrations in plumes from fossil fuel power plants, industrial plants, and petroleum refinery plants have also been studied.[11-13] The excess ozone concentrations have been reported to be 20 to 50 ppb above ambient background concentration several hours downwind from the power plants. As for small-scale industrial and petroleum refinery plumes, excess ozone concentrations of 5 to 30 ppb above ambient background concentrations have been observed.

Ozone Formation Within High Pressure Systems

Apart from the transport of ozone in urban plumes and industrial plumes, elevated ozone concentrations at ground level can also occur in association with stagnating warm high pressure systems. With the passage of a slow-moving warm high pressure system, atmospheric conditions such as relatively high temperatures, high solar radiation intensities, and low wind speeds greatly encourage the production of ozone. Based on studies by Karl,[14] Spicer et al.,[4] and Wolff and Lioy,[15] ozone concentrations associated with warm high pressure systems are estimated to be 70 to 150 ppb. Since the residence time of the air parcels within slow-moving warm high pressure systems are fairly long (e.g., 2 to 6 days), the source of ozone measured at a particular locality is often from areas far away in the upwind direction. Although back trajectory models could be used to reveal possible source regions of ozone formation downwind, the accuracy of these models usually deteriorate rapidly with increased time length for back trajectory reconstruction. In general, the highest ozone levels are associated with the backside of stagnating high pressure systems where air parcels have the longest residence time within the system.[16] On the basis of back trajectories of upper-air wind, Evans et al.[17] concluded that elevated ozone concentrations observed at remote sites of the United States are caused by long range transport of ozone and its precursor over several hundred miles from distant heavily populated regions. In another study by Wolff et al.,[18] air parcels traveling from the midwest and air parcels traveling across New York state from Canada were observed to contain elevated ozone levels of 60 to 130 ppb and 60 to 94 ppb, respectively.

With the presence of warm high pressure systems, the nocturnal radiation inversion is often in place. Similar to the diurnal ozone cycle mentioned above, ozone concentrations at surface locations decrease substantially during the evening hours and into the early morning. During the following morning, the mixing downward of the ozone trapped aloft overnight produces a rapid increase of ozone concentration. At the same time, photochemical ozone formation commences, which in turn generates additional ozone. Therefore, it is not surprising that the highest surface ozone concentrations at a nonurban location are often associated with fumigation by individual or superimposed urban plumes during passage of a warm high pressure system. That is, highly elevated levels of ozone are composites of "natural" background ozone, ozone generated in urban and industrial plumes and ozone formed from precursor accumulation in slow-moving high pressure systems.

DISCUSSION AND CONCLUSIONS

The transport of ozone and its precursors downwind from urban and industrial areas is unquestionable. Urban plumes can extend over hundreds of kilometers downwind and cover thousands of square kilometers of area downwind. In addition, ozone and its precursors remaining aloft are protected from scavenging by the nocturnal inversion. Once aloft, ozone and its precursors can be transported by low-level nocturnal jet stream and upper-level winds far downwind during the nighttime hours.[5] From direct experimental evidence and smog chamber experiment results, it can be seen that ozone plumes can survive for at least two days while traveling over water or traveling aloft.[4] Estimated values of ozone concentrations generated within urban plumes range from 150 to 250 ppb for areas downwind of major population centers.

In addition to ozone plumes from "known" sources, elevated ozone concentrations can also be attributed to ozone formation within slow-moving high pressure systems which have passed over populated areas. Estimated values of this regional ozone concentration range from 70 to 150 ppb for the northeastern United States.

The above review suggests that an additive effect of multiple urban emissions and long range transport can have major impact on the ambient ozone concentrations downwind of the source regions. Clearly, a multiregional control strategy for ozone precursors is needed to solve the ozone problem in New England. Coupled with local strategies in controlling hydrocarbon emissions, a reduction of ozone may be achieved.

A completely different type of strategy in overcoming the noncompliance status in ozone is to adopt a probabilistic "standard" for ozone. Since ozone formation depends heavily on atmospheric conditions (for both in situ ozone formation and transport), a "weather forecast" type of approach may be effective in resolving the noncompliance status currently attained by many states. Apart from atmospheric conditions, considerations should also be given to the geographic position of each city or state. By virtue of their geography, cities and states at the end of a weather system (such as a slow-moving high pressure system) will continuously be affected by their neighbors upwind.

REFERENCES

1. Fishman, J., and P. J. Crutzen. "The Origin of Ozone in the Troposphere," *Nature* 225:855–858 (1978).
2. Viezee, W., W. B. Warren, and H. B. Singh. "Stratospheric Ozone in the Lower Troposphere—II. Assessment of Downward Flux and Ground-Level Impact," *Atmos. Environ.* 17:1979–1993 (1983).
3. Worth, J., and L. A. Ripperton. "Rural Ozone—Sources and Transport," in *Advances in Environmental Science and Engineering,* J. A. Pfafflin and E. N. Ziegler, Eds. (London: Gordon and Breach Science Publishers, 1980), pp. 150-170.
4. Spicer, C. W., D. W. Joseph, P. R. Sticksel, and G. F. Ward. "Ozone Sources and Transport in the Northeastern United States," *Environ. Sci. Technol.* 13:979–985 (1979).
5. Spicer, C. W., G. M. Sverdrup, A. J. Alkezweeny, K. M. Busness, R. C. Easter, and N. C. Possiel. "Ozone Plume Mapping in the New York and Boston Areas," paper presented at the 74th annual meeting of the Air Pollution Control Association, Philadelphia, PA, 21–26 June, 1981.
6. Rubino, R. A., L. Bruckman, and J. Magyar. "Ozone Transport," *J. Air Poll. Control Assoc.* 26:972–975 (1976).
7. Bazzell, C. C., and L. K. Peters. "The Transport of Photochemical Pollutants to the Background Troposphere," *Atmos. Environ.* 15:957–968 (1981).
8. Spicer, C. W., and G. M. Sverdrup. "Tracer Nitrogen Chemistry During the Philadelphia Oxidant Enhancement Study, 1979," Report to the Office of Air Quality Planning and Standards, U.S. Environmental Protection Agency, Research Triangle Park, NC.
9. Sexton, K. "Evidence of an Additive Effect for Ozone Plumes from Small Cities," *Environ. Sci. Technol.* 17:402–407 (1983).
10. Westberg, H. "Ozone in the Combined Baltimore-Washington, DC Plume," EPA/600/S3-85/070 (December 1985).
11. Davis, D. D., G. Smith, and G. Klauber. "Trace Gas Analysis of Power Plant Plumes via Aircraft Measurement: O_3, NO_x and SO_2 Chemistry," *Science* 186:733–736 (1974).

12. Sexton, K., and H. Westberg. "Ambient Hydrocarbon and Ozone Measurements Downwind of a Large Automotive Painting Plant," *Environ. Sci. Technol.* 14:329–332 (1980).
13. Sexton, K., and H. Westberg. "Photochemical Ozone Formation from Petroleum Refinery Emissions," *Atmos. Environ.* 17:467–475 (1983).
14. Karl, T. R. "Ozone Transport in the St. Louis Area," *Atmos. Environ.* 12:1421–1431 (1978).
15. Wolff, G. T., and P. J. Lioy. "Development of an Ozone River Associated with Synoptic Scale Episodes in the Eastern U.S.," *Environ. Sci. Technol.* 14:1257–1260 (1980).
16. Vukovich, F. M., W. D. Bach Jr., B. W. Crissman, and W. J. King. "On the Relationship Between High Ozone in the Rural Surface Layer and High Pressure Systems," *Atmos. Environ.* 11:967–983 (1977).
17. Evans, G., P. Finkelstein, B. Martin, N. Possiel, and M. Graves. "Ozone Measurements from a Network of Remote Sites," *J. Air Poll. Control Assoc.* 33:291–296 (1983).
18. Wolff, G. T., P. J. Lioy, R. E. Meyers, R. T. Cederwall, G. D. Wight, R. E. Pasceri, and R. S. Taylor. "Anatomy of Two Ozone Transport Episodes in the Washington, D.C. to Boston, Mass. Corridor," *Environ. Sci. Technol.* 11:506–510 (1977).

CHAPTER 3

Ambient Levels of Ozone in New England and Interpretation of Trends

Richard P. Burkhart

INTRODUCTION

The Clean Air Act (CAA) requires that all areas of the country comply with the National Ambient Air Quality Standard (NAAQS) for ozone (O_3) by December 31, 1987. The NAAQS for ozone is 0.12 ppm, based on a one-hour average. Any day that has an hourly average over the NAAQS is considered an exceedance day. The CAA allows for three such days in any three year period; more than three exceedance days constitutes a violation of the standard. In 1984, 79.2 million Americans were living in counties that monitored violations of the NAAQS for ozone.[1]

Currently in New England, 45 out of 67 counties[2] are not in attainment for ozone, meaning the ozone standard has recently been violated. These nonattainment areas include the entire states of Connecticut, Massachusetts, and Rhode Island. In the past five years (1982–1986) ozone concentrations in excess of twice the standard have been measured in Connecticut, Rhode Island, and Massachusetts.

Since elevated ozone levels can cause adverse effects to human health, forests, and crops,[3,4] these excessive ozone levels must be reduced. The CAA mandates an ozone reduction to the NAAQS by December 31, 1987, and progress has been made to reduce ozone levels throughout the country.

Since ozone is a secondary pollutant, meaning it is not emitted directly from sources, ozone reduction is only possible by reducing emissions of the pollutants that react to form ozone, the so-called precursor pollutants: oxides of nitrogen

(NO$_x$) and nonmethane hydrocarbons (NMHC). The EPA estimates that volatile organic compound (VOC) emissions have been reduced by 45% in Massachusetts and Connecticut (the two largest emitters in New England) from 1980 to 1987.[5] Unfortunately, monitored ozone levels have not shown as dramatic a decrease. In fact, ozone levels in New England increased from 1981 to 1983, and since 1983 have decreased to only slightly below 1981 levels. The reason for this apparent discrepancy is due to the nonlinear chemistry of ozone production and its dependence on meteorology. The solution to the ozone dilemma is the reduction of precursor emissions, but determining the quantity of necessary reductions is complex.

Both chemistry and meteorology influence ozone formation and both must be accounted for in any analysis or modeling of ozone. This chapter will address these influences in order to give the reader a better understanding of the complexity of ozone formation. The chapter explores the monitored ozone data in New England for the past seven years and analyzes the existence of discernible trends. Basic ozone modeling techniques are discussed, including the Empirical Kinetic Modeling Approach (EKMA), regional oxidant modeling, and trajectory analysis. The chapter concludes with a discussion on ozone trends in New England and recommendations on future ozone monitoring and modeling.

OZONE FORMATION AND TRANSPORT

Ozone formation at the earth's surface requires precursor pollutants and meteorological conditions that are readily available during New England summers. Warm New England summer winds, which blow predominantly from the southwest, advect large amounts of the precursor ingredients necessary for ozone formation from New York and New Jersey into New England. The summer sun "cooks" these precursors into ozone while they are advecting into New England. This section describes the basic chemistry of ozone formation, and discusses the meteorology conducive to ozone formation.

Basic Ozone Chemistry

Ozone, at ground level, is formed by photochemical reactions between oxides of nitrogen (NO$_x$), hydrocarbons (HC), carbon monoxide (CO), and oxygen (O$_2$). The chemical reactions are complex, and detailed descriptions are available elsewhere.[6-10] Simply stated, ozone formation can be thought of in the following manner:

$$CO + NO_x + HC + O_2 + SUN => O_3 \qquad (1)$$

where:

SUN — Strong ultraviolet light.

This chemical formula states that carbon monoxide, oxides of nitrogen, hydrocarbons, and oxygen react in the presence of sunlight (photochemical reaction) to form ozone. The reactivity of this reaction is dependent on several parameters, including the reactivity of the hydrocarbons,[6,8,9] insolation,[10] and temperature.[8,11] The reactivity of the hydrocarbons is a very important parameter in the ozone formula. In typical urban and rural ozone formation, carbon monoxide and some hydrocarbons, most notably methane, are not as conducive to ozone formation as are some other hydrocarbons. These more ozone-conducive hydrocarbons are known as volatile hydrocarbons, volatile organic compounds (VOC), and non-methane hydrocarbons (NMHC). [The term VOC in this chapter refers to emissions of volatile hydrocarbons, while the term NMHC refers to ambient concentrations of volatile hydrocarbons.] Because of the varied reactivities of hydrocarbons, and the extremely low reactivity of methane, methane is not an input to most ozone models, including EKMA. Carbon monoxide is also not an explicit input to EKMA, but it is still included in the reaction chemistry. On a global scale, however, carbon monoxide and methane are both important to the ozone budget.[6]

Since carbon monoxide, methane, and other nonvolatile hydrocarbon emissions are not as important as the emissions of the more volatile hydrocarbons in the rural/urban ozone formation that dominates the New England ozone problem, a simplified ozone formula can be used:

$$NO_x + VOC + SUN => O_3 \qquad (2)$$

The actual chain of chemical reactions which forms ozone is highly nonlinear and full of ozone source and sink reactions. The nonlinearity of the reactions means that if one of the ingredients on the left hand side of Equation 2 is reduced by a fixed amount, say $X\%$, the ozone concentration will not change by $X\%$, but will change by some other amount, $Y\%$, and this change may represent an increase or a decrease in ozone concentration. For example, the 1982 State Implementation Plan for Massachusetts[12] shows that an 8% decrease in NO_x emissions and a 26% decrease in VOC emissions would be required in Boston to reduce ambient ozone levels in Georgetown, Massachusetts by 7%. This analysis was performed using the EKMA model. EKMA is discussed below.

The source and sink reactions affecting ozone are also complex. In some circumstances, an increase in NO_x emissions can cause a decrease in ozone locally, because NO_x initially acts as an ozone sink (nitric oxide), reducing ozone levels.

[NO_x is composed of nitric oxide (NO) and nitrogen dioxide (NO_2). Most stationary and mobile sources emit NO which is an ozone sink (scavenger). Additional photochemistry converts NO to NO_2. This NO_2 is an ozone source (precursor), and causes increased ozone levels downwind of large NO_x sources, such as urban areas. Because of the scavenging effect of NO, Center Business Districts (CBD) of large cities often have lower ozone levels than their surrounding suburbs.] Nitric oxide, however, is soon converted to nitrogen dioxide, an ozone source, and thus elevates ozone levels downwind. Because of this complex relationship, NO_x control alone is not considered by the Environmental Protection Agency to be an effective means for attaining the NAAQS for ozone, unless a photochemical grid modeling study is performed.[13] Photochemical grid models are discussed further below. A study by Dodge[14] states that "NO_x control by itself, has a detrimental impact on ozone levels near source areas" and "NO_x control therefore could result in greater population exposure to high ozone levels than would be the case if only VOC is reduced."

Ozone Conducive Meteorology

During the summer, differential heating between the North American continent and the Atlantic Ocean causes a semipermanent high pressure area (both aloft and at the surface) to form in the vicinity of Bermuda, and it has become widely known as the Bermuda high. The Bermuda high conjures up thoughts of hot, humid, semitropical air. Unfortunately for New Englanders this air is often laced with unhealthful levels of ozone. The dynamics of the Bermuda high, in conjunction with emissions of pollutants from the large cities along the Atlantic seaboard, contribute to these unhealthful ozone levels.

Because of the dynamics of the atmosphere, there are light, clockwise moving surface winds around high pressure areas. This clockwise movement around highs causes New England's surface winds to be southwesterly while under the influence of the Bermuda high. These southwest winds bring in pollutants from upwind areas, such as New York and New Jersey, into New England. The flow of pollutants from these huge source areas can be thought of as large rivers of smog, called plumes. These smog plumes can be traced for hundreds of miles.[10,15-18] Frictional forces and dynamics cause the surface winds of high pressure areas to diverge and produce downward moving air (subsidence), which stabilizes the atmosphere and reduces vertical mixing. The reduced vertical mixing forces pollutants to remain in a rather thin layer of the atmosphere known as the boundary layer. The subsidence also causes a minimum in cloud cover which greatly aids in the photochemical formation of ozone. The warm temperatures of summer also aid in the formation of ozone. The role of temperature on ozone formation is not totally understood, but it is believed to help increase NMHC reactions. Smog chamber studies of ozone formation have shown increased reaction rates with increased temperatures.[8,11]

Sea Breeze Influence on Ozone

Another interesting meteorological phenomenon that is hypothesized to contribute to elevated ozone levels in coastal New Hampshire and Maine is the sea breeze effect. At certain times during the summer, meteorological conditions will cause Boston's VOC and NO_x emissions (the Boston plume) to be transported toward the north-northeast, traveling out over Cape Ann, Massachusetts, and toward the Maine coast. This type of flow is often associated with the Bermuda high. The Boston plume traveling over open ocean accomplishes several things which promote ozone formation and minimize ozone depletion. First, the cooler ocean waters suppress mixing, which causes the plume to remain vertically thin, and thus concentrated. Second, decreased vertical mixing also inhibits plume contact with the surface, a known ozone sink. Third, the more stable ocean air further inhibits cloud growth already reduced by subsidence. Clouds, if present, would increase mixing. The lack of clouds allows full sun to drive the photochemical ozone reaction.

As the day progresses and the land warms, air begins to rise over the land and the cooler, denser ocean air begins to flow landward, and the well-known sea breeze becomes established. As the cooler sea breeze, laden with its ozone plume, encounters the warmer and aerodynamically rougher land, mixing occurs and the ozone plume is rapidly mixed to the ground, and the coastal areas experience rapid increases in ozone. Ozone violations have been reported as far north as Acadia National Park (200 miles from Boston), and are more frequent at Cape Elizabeth, Maine, and Portsmouth, New Hampshire. Since sea breeze fronts (the leading edge of the sea breeze) are parallel to the coastline, the sea breeze begins at similar times along entire sections of coastline. Reported violations at coastal sites in Maine during sea breeze events generally occur within one hour of each other,[19] supporting the theory that ozone violations in Maine are the result of over-water transport of the Boston plume. Several ozone studies also help support this theory.[20,21]

Long-Range Transport of Ozone

The theory of ozone exceedances in Maine being caused by the Boston plume rests on the assumption that the plume can remain a concentrated, definable plume for hundreds of miles. As mentioned previously, several studies have traced ozone plumes over 100 miles. Several investigations have also tracked the Boston ozone plume using airplanes, and verified that it can also be traced for at least 100 miles. Figure 1 shows the results of two of these studies.[16,18] According to Spicer et al.,[16] the data in Figure 1 (bold font) show that the high ozone concentrations in the plume upwind of Boston originated in the Philadelphia, New York, and New Jersey areas the day before and were augmented by fresh precursor emissions from Boston on the day studied. These new pollutants formed ozone which

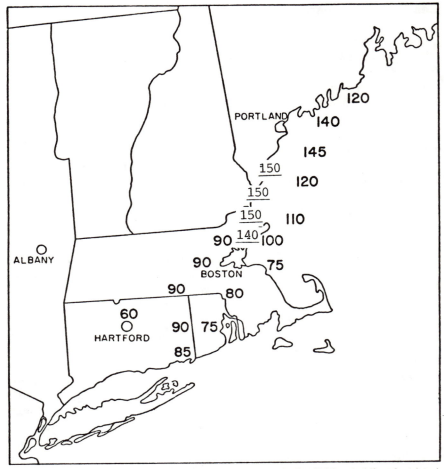

Figure 1.   Ozone concentrations for July 24, 1975 at an altitude of 300 m, boldface [reprinted with permission from C. W. Spicer et al., "Ozone Sources and Transport in the Northeastern United States," *Environ. Sci. Technol.* 13(3):975–985, 1979, American Chemical Society[16]] and for August 1, 1980 at an altitude of 760 m, underlined numbers (from NECRMP[18]).

was tracked northward toward Maine. A similar explanation is feasible for the Northeast Corridor Regional Modeling Project (NECRMP) data shown in Figure 1, (underlined, narrow font). These data show ozone values downwind of Boston were measured to be over 120 ppb.[18] Thus, the Boston plume does exist and the centerline concentrations can exceed the NAAQS for ozone.

This long-range transport of ozone is not only important in explaining violations of the ozone standard in Maine, but is also a major issue for southern New England. Long-range transport, as shown in Figure 1, also adds to the ozone burden in southern New England, and helps contribute to violations of the ozone standard.

Other Sources of Ozone

Another source of ozone at ground-level that is not unique to New England is the stratosphere. The lower stratosphere above New England contains concentrations of ozone that average about 3 ppm.[22] It is this stratospheric ozone that has become known as the "ozone layer." Ozone in the stratosphere is natural and beneficial to life on earth. This natural ozone source is sometimes tapped by atmospheric processes, which causes stratospheric air to mix with tropospheric air. There are seasonal variations in the rate at which stratospheric air is mixed into the troposphere[6,23] and therefore, there is a seasonality to the stratospheric contribution to natural background levels of tropospheric ozone.

A study conducted by Singh et al.[23] reported that ground-level background ozone values averaged about 30 ppb, and nearly all of this is of stratospheric origin. They reported yearly maximums, as high as 80 ppb, in the late winter and early spring, and minimums in the fall, as low as 15 ppb. Altshuller[10] in contrast to Singh et al.[23] states that typical stratospheric contributions to ground-level ozone are on the order of 10 ppb, and the remaining portion is from natural precursor emissions.

In addition to a slow exchange of stratospheric and tropospheric air that may lead to elevated ozone levels at ground-level, strong cold fronts, especially those during the springtime, can inject large quantities of stratospheric air directly into the troposphere.[24-26] Such injections of stratospheric air are called stratospheric intrusions. Stratospheric intrusions of ozone may cause high ground-level ozone concentrations. These intrusions, in combination with rain and thundershower activity which often accompany strong cold fronts, have reportedly caused violations of the NAAQS for ozone. The role of shower activity in this process is not completely understood, but the showers probably aid in mixing the ozone aloft to the ground. A study by Lamb[23] showed a one-hour ozone concentration of 0.23 ppm from such an event. The actual effect of stratospheric intrusions on ground-level concentrations of ozone is still controversial. Altshuller[10] states that "stratospheric O_3 contributions to surface locations exceeding 100 ppb may be questionable."

The stratosphere is not the only natural source of ground-level ozone. Biogenic emissions of precursor pollutants (hydrocarbons, carbon monoxide and oxides of nitrogen)[6] also contribute to ground-level ozone. Trainer[27] cites biogenic emissions of hydrocarbons, especially isoprene, from conifers as substantial contributors to the natural background concentrations of ozone. Linvill et al.[28] implicates soil and actively growing green plants as major sources of nitrogen oxides, an ozone precursor.

Studies of the ozone levels in pristine areas of the globe far from the effects of any possible anthropogenic sources of precursor pollutants, and historic data from the nineteenth century are often used to estimate the concentrations of natural ground-level ozone. Logan[6] uses recent data from pristine areas in Oregon and Montana and historical (preautomobile) ozone data from Michigan, to estimate the natural ground-level ozone concentration at mid-latitudes. The Michigan ozone data[6,28] from 1876 to 1880 (measured using Schoenbein paper) show an average daily maximum in April of over 60 ppb and an average daily minimum in December of approximately 20 ppb. The absolute value of these readings may be slightly biased because of the method used,[29] but the seasonal fluctuations are thought to be accurate. Recent data from Oregon and Montana show similar results. Historic data from Vienna and Klagenfurt, Austria, collected from 1854 to 1873, using Schoenbein paper, show seasonal trends similar to the Michigan data, with maximums in the spring and minimums in the fall.[29] The spring maximums at Klagenfurt are, on average, higher than those at Vienna. Klagenfurt is located 300 km from Vienna in a high mountain valley. Elevation is thought to be a factor in ozone concentrations, with high elevations receiving higher ozone doses than lower elevations.[30]

Altshuller[10] states that average ground-level ozone values at pristine mid-latitude locations are on the order of 30 to 50 ppb and have seasonal variations, and pristine areas in the tropics have concentrations on the order of only 15 to 20 ppb. The low values in the tropics may be due to the location of the monitoring sites. Logan[6] shows that coastal monitors in Florida, Louisiana, and Japan have summer minimums in ozone in contrast to inland sites that have summer maximums. Logan[6] attributes these minimums to exceptionally clean maritime air that is advected over the monitors by the diurnal sea breeze. If the tropical monitoring sites used by Altshuller[10] are located close to the shoreline, they may be overly influenced by clean sea breeze circulations. The maritime air is low in ozone because the oceans, especially the south Pacific, are thought to be large sinks for ground-level ozone.

Correlation of Meteorological Variables to Elevated Ozone Levels

Several studies have shown correlations between meteorological variables and elevated ozone levels. Warm temperatures have shown positive correlation with ozone levels.[31-33] Other meteorological variables such as wind speed, wind direction, insolation,[31-33] precipitation,[33] and visibility[32] have all been correlated with elevated ozone levels. Reference 34 provides a good overview of the studies that correlate meteorological variables to high ozone episodes.

In light of these previous studies, this chapter explores the relationship between ozone exceedance days in New England and specific meteorological conditions. An exceedance day is any day in which one of the ozone monitoring sites in New England (Fig. 2) records an exceedance of the NAAQS for ozone. The site names for the sites shown in Figure 2 are listed in Table 1.

A previous study[19] showed a strong relationship between exceedance days and maximum temperatures greater than or equal to 86°F as recorded at Bradley

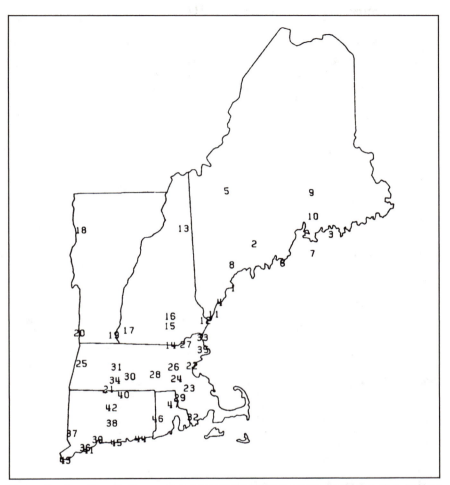

Figure 2. Location of all current and selected historical ozone monitoring sites in New England (selected sites monitored ozone for at least three ozone seasons since 1980). All sites are designated by Arabic numerals. The site numbers are related to the site names listed in Table 1.

International Airport (BDL), north of Hartford, Connecticut. BDL was chosen as the meteorological station for the following reasons:

- noncoastal location
- local climatological data summaries (LCDS) are available
- worst New England ozone is in Connecticut
- elevation is close to sea level
- rural location

Table 1. List of Sites Shown in Figure 2.

Site #	Site Name	Data Record	Used in 17 Site Study	Used in 30 Site Study	Site Moved
1	Cape Elizabeth, ME	80–86	yes	yes	1981
2	Gardner, ME	80–86	yes	yes	
3	Acadia, ME	83–86			
4	Kennebunkport, ME	82–86			1982
5	Sugarloaf, ME	86			
6	Port Clyde, ME	86			
7	Isle au Haut, ME	86			
8	Cumberland, ME	80–83			
9	Penobscot, ME	80–82			
10	Hancock, ME	85–86			
11	Cape Neddick, ME	86			
12	Portsmouth, NH	80–86	yes	yes	
13	Berlin, NH	81–86		yes	
14	Nashua, NH	81–86		yes	1984
15	Manchester, NH	80–86		yes	1984
16	Pembroke, NH	85–86			
17	Keene, NH	81–83			
18	Burlington, VT	80–86	yes	yes	
19	Brattleboro, VT	82–85		yes	
20	Bennington, VT	86			
21	Agawam, MA	80–86	yes	yes	1982
22	East Boston, MA	80–86	yes	yes	1984
23	Easton, MA	80–86	yes	yes	
24	Medfield, MA	80–85	yes	yes	
25	Pittsfield, MA	80–85	yes	yes	1983
26	Sudbury, MA	80–86		yes	
27	Lawrence, MA	81–86		yes	
28	Worcestor, MA	80–86		yes	
29	Attleboro, MA	80–85		yes	1982
30	Ware, MA	80–86		yes	
31	Amherst, MA	83–86			
32	Fairhaven, MA	82–86			
33	Plum Island, MA	83–86			
34	Chicopee, MA	83–86			
35	Hamilton, MA	80–82			
36	Bridgeport, CT	80–86	yes	yes	
37	Danbury, CT	80–86	yes	yes	
38	Middletown, CT	80–86	yes	yes	
39	New Haven, CT	80–86	yes	yes	
40	Stafford, CT	80–86	yes	yes	
41	Stratford, CT	80–86	yes	yes	
42	East Hartford, CT	81–86		yes	
43	Greenwich, CT	81–86		yes	
44	Groton, CT	81–86		yes	1985
45	Madison, CT	81–82 84–86		yes	
46	Kent, RI	80–86	yes	yes	
47	Providence, RI	80–86	yes	yes	

The previous study[19] used only data from 1983 to 1986. This chapter expands the scope of the previous study to include data from 1980 to 1986, and also examines the relationship between high ozone levels and two additional meteorological variables: average daily wind speed and percentage of possible sunshine. Both variables are available on the LCDS for BDL (Hartford).[35]

Table 2 shows the number of exceedance days for the years 1980 to 1986, and the number of days that the maximum temperature at BDL exceeded 86°F and 90°F. Both parameters are well correlated to the number of exceedance days with correlation coefficients of 0.90 and 0.87, respectively. A further analysis of the ozone and temperature data stratified into frequency distributions for exceedance and nonexceedance days (Table 3), for the months of May to September 1980–1986, allows one to compute the biserial correlation,[36] between ozone exceedance days, nonexceedance days, and maximum temperature. Only the warmer months are used in this study, since rarely do ozone exceedances occur in the cooler months of the year, and these cooler months would unduly bias the sample.

Table 2. **Highest Second-Maximum Ozone Concentration and Number of Exceedance Days in Region I, and the Number of Days the Maximum Temperature at Bradley International Airport was Greater Than or Equal to 86°F and 90°F for the Period 1980 to 1986.**

Year	1980	1981	1982	1983	1984	1985	1986
Highest Second-Maximum Ozone Concentration[a] (ppb)	276	202	230	240	224	168	160
Ozone Exceedances (days)	55	35	38	65	43	29	16
Max. Temperature ≥86°F (days)	43	32	27	58	38	25	27
Max. Temperature ≥90°F (days)	19	13	11	38	12	5	7

[a]Every highest second-maximum concentration occurred in Connecticut.

Table 3. **Dichotomous Frequency Distribution Between Maximum Temperatures at the Bradley International Airport and Exceedance or Non-Exceedance Days in Region I for the Period May to September, and the Years 1980 to 1986.**

Maximum Temperature (°F)	No. of Exceedance Days	No. of Non-Exceedance Days	Total
50.0–54.9	0	12	12
55.0–59.9	0	21	21
60.0–64.9	0	57	57
65.0–69.9	0	97	97
70.0–74.9	4	146	150
75.0–79.9	12	180	192
80.0–84.9	53	186	239
85.0–89.9	110	87	197
90.0–94.9	73	10	83
95.0–99.9	23	0	23
Total	275	796	1071

The biserial correlation between maximum temperatures and ozone exceedance days is 0.78. The mean maximum temperature of exceedance days is 88°F, and the mean maximum temperature of nonexceedance days is 76°F. These two means are significantly different at the 99% confidence level.

The number of days with maximum temperatures above specified levels as recorded at BDL is also correlated with the yearly highest second-maximum ozone value (a common air quality measure) as measured in New England for the years 1980 to 1986 (Table 2). Days with maximum temperatures greater than or equal to 86°F have a correlation coefficient of 0.66 when correlated to the highest second-maximum ozone value, and days with maximum temperature greater than or equal to 90°F have a slightly smaller correlation of 0.61. These correlations are less than their respective correlations with the number of exceedance days discussed above. The correlation of maximum temperatures to both exceedance days and the highest second-maximum monitored ozone concentrations show that temperature is positively correlated with ozone concentrations.

The daily average wind speed data and ozone exceedance days (Table 4) show a biserial correlation of −0.16, with mean wind speeds of 6.3 mph and 7.0 mph for exceedance and nonexceedance days, respectively. These means are significantly different at the 99% confidence level. The negative biserial correlation, although small, shows that high winds are not as conducive to ozone formation as less windy conditions. This makes sense, since high winds aid in dilution. The small biserial correlation is also expected, since a previous study[33] stated that high wind speeds are not as important in the dilution of rural ozone as they are in the dilution of urban ozone, and rural sites are much more likely to monitor ozone exceedances.

The percentage of possible sunshine is used as a measure of insolation. Its frequency distribution is shown in Table 5. These data are not normally distributed

Table 4.   Dichotomous Frequency Distribution Between Mean Wind Speed at the Bradley International Airport and Exceedance or Non-Exceedance Days in Region I for the Period May to September, and the Years 1980 to 1986.

Mean Wind Speed (mph)	No. of Exceedance Days	No. of Non-Exceedance Days	Total
0.0– 1.9	0	1	1
2.0– 3.9	30	63	93
4.0– 5.9	102	257	359
6.0– 7.9	86	234	320
8.0– 9.9	44	140	184
10.0–11.9	12	73	85
12.0–13.9	1	19	20
14.0–15.9	0	9	9
Total	275	796	1071

Table 5. Dichotomous Frequency Distribution Between Percentage of Possible Sunshine at the Bradley International Airport and Exceedance or Non-Exceedance Days in Region I for the Period May to September, and the Years 1980 to 1986.

Possible Sunshine (%)	No. of Exceedance Days	No. of Non-Exceedance Days	Total
90.1–100	69	207	276
80.1–90	62	89	151
70.1–80	41	78	119
60.1–70	32	59	91
50.1–60	26	64	90
40.1–50	26	49	75
30.1–40	11	42	53
20.1–30	2	44	46
10.1–20	3	44	47
0.1–10	2	57	59
0.0	1	63	64
Total	275	796	1071

so a significance test is not performed. The mean percentage of possible sunshine is higher on exceedance days than on nonexceedance days, 73% versus 58%, and there is a slight positive biserial correlation, 0.28, between percentage of possible sunshine and ozone exceedance days. Note, however, that when there is 100% available sunshine, it is more likely not to have an ozone exceedance (95 out of 119 days), than to have one (24 out of 119 days). A likely explanation for this apparent discrepancy is based on the cleansing effect of summer cold fronts. Most New England summer cold fronts are accompanied by a wind shift to the northwest which advects cooler, dryer, and cleaner air from the Adirondack and Green Mountains over all of southern New England. The weather following a summer cold front is often cloud-free, allowing 100% possible sunshine. The lack of anthropogenic precursor pollutants in the postfrontal air mass greatly reduces ozone potential, and ozone exceedances are rarely measured. An examination of the meteorological data for BDL from 1980 to 1986 show that on six of the nine days with maximum temperatures greater than 90°F and no monitored ozone exceedances, the winds were northwesterly.

Ground-level insolation can also be affected by ozone concentrations. Horie et al.[32] shows that visibility and high ozone are inversely correlated, so when ground-level ozone concentrations are high, visibility is low due to increased haze. The haze is the result of increased photochemical oxidants. Thus, on days with high ozone levels, ground-level insolation is lower than it would be in the absence of pollution. Percentage of possible sunshine is not generally affected by haze, but total insolation, when measured, would be affected. Total insolation might show a different correlation to ozone than percentage of possible sunshine.

Total insolation data are not available via the LCDS, and therefore have not been correlated to ozone levels in this study.

Returning to the relationship between percentage of possible sunshine and ozone exceedance days, the data clearly show that when percentage of possible sunshine is low, ozone exceedances are rare. For percentage of possible sunshine less than or equal to 40% there were only 19 exceedance days out of a possible 269 total days, or less than 9% of those days had exceedances. On days with 30% or less possible sunshine, the percentage of days with exceedances is less than 4%. To summarize, cloudy days are not conducive to ozone formation.

OZONE LEVELS IN NEW ENGLAND

Ozone is currently monitored at 39 sites in New England. The ozone monitors measure ozone during only the ozone-conducive time of the year from April to October. This period is known as the ozone season, with July being the month in New England with the highest number of exceedance days. July has averaged 14.3 exceedance days over the past seven years.

Figure 2 shows the location of the ozone monitors in New England. The figure includes all current ozone monitoring sites, as well as any site that monitored for a minimum of three ozone seasons since 1980. Table 1 lists the names and the period of data record for each site shown in Figure 2, and lists which sites have been moved and which sites are used in the 17 and 30 site trend analysis discussed below.

Ozone monitoring sites can be divided into two main categories: urban sites, located within the Central Business District (CBD) of major urban areas such as Providence, Boston, and Hartford, and suburban/rural sites located in areas surrounding the cities. Because of nitric oxides scavenging ozone in the CBD, the suburban and rural sites often measure higher ozone levels than their urban counterparts. The CBD sites are generally the sites that also measure the precursor pollutants, a necessary input to ozone models, such as EKMA.

This section discusses the ozone levels measured in New England over seven years (1980–1986). The data prior to 1980 are not used because of a change in the technique used to calibrate the ozone monitors which occurred in the 1978 to 1979 timeframe. It is difficult to account for this calibration change in data analysis.[1]

Trends in the Second-Maximum Ozone Value

As discussed previously, high ozone levels are correlated with high temperatures. Thus in years with very warm summers, such as 1983, ozone levels are high, while during cooler summers the ozone levels may be lower. Therefore, when looking at trends in New England ozone levels, one must be cognizant of the meteorological conditions that prevailed during the period of record.

Another factor affecting ozone concentrations is the emission of the precursor pollutants, VOC and NOx. Because VOC emissions have decreased in New England since 1980, and continued decreases are expected, the trend in ozone levels should reflect these decreases. Unfortunately, since ozone chemistry is highly nonlinear and dependent on meteorology, any effect of decreased precursor emissions on ozone levels may be masked.

Figures 3 and 4 show box and whisker plots of the second-maximum ozone level measured at 17 sites (Fig. 3) and 30 sites (Fig. 4) in New England for the past seven years. Table 1 shows which sites were used in both the 17- and 30-site samples. The reason both 17- and 30-site samples are shown is because ozone sites are not permanent, and a site may be moved for various reasons. Table 1 lists several sites that were moved during the period. None of these sites were moved more than 10 miles, and these moves should not have affected the data quality. The 17 sites of Figure 3 have monitored ozone data since 1980. Two of these sites, unfortunately, ceased operation in 1986. Figure 4 is nominally based upon 30 sites; however, some sites were moved from one year to the next (see Table 1), so it is not as stable a sample as the 17 sites used in Figure 3. Both figures are included to show the reader that there is not a great difference between a small, stable 17-site sample of the ozone data and a larger, more inclusive

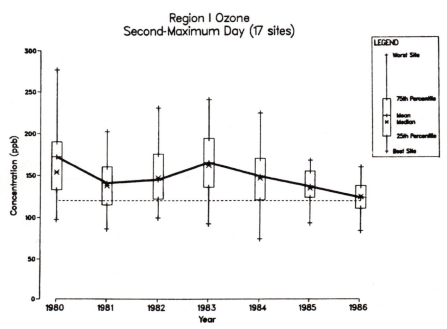

Figure 3. Box and whisker plot of the Region I ozone second-maximum day for the years 1980 to 1986 for the 17 sites named in Table 1.

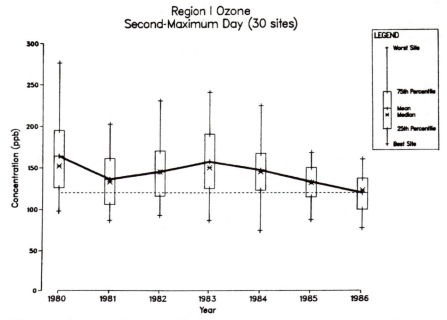

Figure 4.   Box and whisker plot of the Region I ozone second-maximum day for the years 1980 to 1986 for the 30 sites named in Table 1.

30-site sample. However, because of the consistency of sites used in the 17-site sample, only these data will be used in all future discussions.

Figure 3 shows that both the mean and median second-maximum daily ozone concentrations in New England have decreased steadily since 1983. A comparison of the mean second-maximum concentration for the year 1983 (165 ppb), with the mean second-maximum for 1986 (124 ppb), shows a decrease of 41 ppb. This decrease is significant at the 99% confidence level. Remember, though, that 1983 had a very warm summer (38 days at BDL were at or above 90°F, compared to the 31 year average of 18), while 1986 had a very cool summer (7 days at BDL were at or above 90°F). Comparing the ozone data for 1981 and 1984, two years with similar summertime conditions (10 and 12 daytime highs at or above 90°F, respectively) allows for a better comparison of ozone trends. The mean second-maximum ozone concentration rose 8 ppb from 1981 to 1984, but this is not significant at the 99% confidence level. Individual second-maximum values at monitoring sites in New England had differences ranging from −20 ppb to +53 ppb for this same time period. Such large differences from year to year are quite common, even between years with similar meteorology, such as 1981 and 1984.

Trends in Number of Ozone Exceedance Days

Another statistic that can be used in ozone trend analysis is the number of exceedance days that occur each year. An exceedance day for New England (EPA Region I) is any day that one or more ozone monitors in New England measures an ozone value greater than the NAAQS. A state exceedance day is any day one or more of the ozone monitors in a state records an exceedance of the NAAQS.

Figure 5 shows the trend in exceedance days for the past four years for the six states in EPA Region I, and the region as a whole. Figure 5 clearly shows that the number of exceedance days in Region I is declining. The data also show that 1986 had the fewest exceedance days for five out of the six states in the past four years, and in fact 1986 had the fewest ozone exceedance days since at least 1980 (not shown). Remember, however, that 1986 was one of the coolest and cloudiest summers on record in New England.

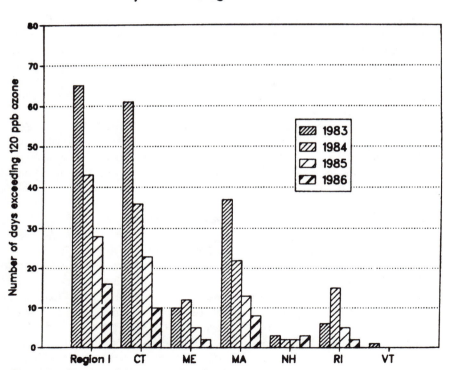

Figure 5. Number of days exceeding the ozone standard for the years 1980 to 1986 for each of the six states in Region I and Region I as a whole.

Discussion on Ozone Trends in New England

Both measures of ozone levels in New England, mean second-maximum and number of exceedance days, have shown downward trends since 1983, and these downward trends are encouraging. Precursor emissions of VOC have also shown downward trends, and this trend is expected to continue. However, the real significance of the downward trend in ambient ozone in New England is suspect, because of the correlation between high ozone and warm temperatures. Indeed, identification of meaningful trends in ozone analysis may not be possible with the limited amount of ozone data available, and the compounding problem of non-linear chemistry, variable meteorology, and the continued movement of long-term ozone monitoring sites. The 1987 ozone season may allow us to better understand the true trend in New England ozone levels.

OZONE MODELING IN NEW ENGLAND

In order to reduce ground-level ozone concentrations below the NAAQS, the ozone precursor pollutants, especially VOC, must be reduced. The amount of VOC reduction necessary depends on several variables, such as current ozone levels, the meteorology of an area, and the amount of long-range ozone and precursor transport affecting the area. These variables are generally input into ozone predictive models in order to calculate the size of the VOC emission reduction necessary to achieve ozone attainment.

Two types of ozone predictive models are generally acceptable for regulatory modeling: city-specific EKMA and regional ozone modeling. Regional ozone models include the Urban Airshed Model and the Regional Oxidant Model (ROM). A third type of model, not specific to ozone but often used to better understand ozone violations and transport, is an atmospheric trajectory model.

The Empirical Kinetic Modeling Approach (EKMA)

The EKMA model is the most widely used ozone predictive model. [The EKMA model as used in this chapter includes the Ozone Isopleth Plotting Package (OZIPP).[37]] The EKMA model is based on chemical relationships between precursor emissions and final ozone concentrations, established using smog chambers. The smog chamber studies were accomplished by filling smog chambers with known concentrations of precursor emissions (VOC and NO_x) and exposing these chambers to ultraviolet radiation. The ultraviolet radiation caused the precursor pollutants to react and form ozone, which was monitored as the experiment progressed. These smog chamber studies allowed the researchers to produce a series of empirical curves which relate precursor reductions to ozone reductions. These empirical curves are known as ozone isopleths.[37]

These empirical isopleth diagrams are at the heart of the EKMA approach. The isopleths used in EKMA are based upon inputs to the EKMA model. These inputs to EKMA include:

- VOC and NO_x emission densities
- Ambient NMHC and NO_x concentrations
- Meteorological data
- Estimates of ozone transport.

The biggest advantage of EKMA is its simplicity. EKMA executes quickly on a mainframe computer, with a computer cost of less than $100. EKMA's biggest disadvantage is its poor treatment of ozone transport, and its one city/one emission reduction mode. EKMA cannot be effectively used in states, such as Connecticut, that are overwhelmed by ozone transport from distant source regions. EKMA usually is not used to predict ozone levels, but it is used to predict the emission reductions necessary to reduce ozone concentrations at a monitored receptor region; therefore, ozone should be monitored in an area before EKMA is used. This monitoring requirement is extremely important, and it demands a prudent, extremely high quality and high data capture rate monitoring effort, in order to ensure that EKMA will predict the correct emission reductions necessary to achieve attainment.

Three-Dimensional Photochemical Models

The Urban Airshed Model[38] is a more sophisticated ozone model than is EKMA. Airshed can be used to predict ozone concentrations at select receptor points. Airshed uses a fully three-dimensional grid to predict ozone formation and to account for local meteorology and long-range transport. Airshed has a good chemistry module, treats atmospheric diffusion directly, and even accounts for surface removal of ozone, via ozone sinks.

Two of the biggest disadvantages of Airshed are that it requires detailed meteorological input, both at the surface and aloft. This level of detail is not routinely available in New England, and would only be available during a major air quality study. Such a detailed study was performed in 1980 in the greater New York metropolitan area. That study is a part of the NECRMP study and is entitled "Oxidant Modeling for the New York Metropolitan Area Project (OMNYMAP)." Airshed has been run based on the OMNYMAP data, and a final report will soon be available.[39] Another disadvantage to Airshed is that it requires detailed precursor emission data. Emission data are extremely difficult to inventory. Airshed is also complicated to initialize and execute, and the output is difficult to interpret. The model requires a substantial amount of computer time on a large mainframe. A typical computer run time charge for one Airshed execution on the OMNYMAP dataset is $5000.[40]

In order to overcome some of the limitations in Airshed and to formulate a better three-dimensional photochemical model for ozone, the EPA developed the Regional Oxidant Model (ROM).[41] ROM is a fully three-dimensional photochemical gridpoint model that can simulate hourly average ozone and precursor concentrations over the entire Northeastern United States. ROM, unlike Airshed, uses routinely available meteorological data as input, but like Airshed, it still requires detailed emission data and substantial computer resources. A typical ROM simulation for the Northeast costs over $50,000 for computer time alone.[40] Such high computer costs are a concern to states that need a model like ROM in order to fully address the complex chemistry, meteorology, and transport of ozone and its precursors.

Trajectory Modeling

Another type of model used to analyze the causes of high ozone levels in New England is the atmospheric trajectory model. Two popular atmospheric trajectory models used in New England are the Atmospheric Trajectory and Diffusion Model (ATAD)[42] and the Branching Atmospheric Trajectory Model (BAT).[43] Both models use wind and temperature data collected at the standard upper-level meteorological stations as input. The wind data from these stations, when used in either BAT or ATAD, are first averaged in the vertical, then interpolated in the horizontal using a distance-weighted averaging scheme. Since upper-air data are generally available only every 12 hours, and both the BAT and ATAD model require wind data every 6 hours, the data must be linearly interpolated in time as well as space. These temporal and spatial interpolation schemes increase the uncertainty of the predicted trajectories.

Both ATAD and BAT when applied to ozone analysis are generally used in the "backward" trajectory mode, which means that the path of pollutants affecting a region, say Portland, Maine, is tracked backward in order to discover where the pollutants originated. Several studies have used either BAT or ATAD to study ozone exceedances in Maine in order to discover the source of those exceedances.[19,44,45] Figure 6 shows an example of the output of the BAT model used to study an ozone exceedance at Cape Elizabeth, Maine, near Portland, on June 11, 1984.[19] The highest hourly average ozone concentration at the Cape Elizabeth monitor on June 11, 1984 was 133 ppb. The BAT model output is overlaid on a map background of the Northeast. The data show that the trajectory arriving in Maine on June 11, 1984 originated 60 hours earlier over western Pennsylvania and Ohio, and traveled over southern Vermont and extreme northwestern Massachusetts before reaching Maine. The branches in the trajectory are the result of vertical wind shear.

Atmospheric trajectory models such as ATAD and BAT can be useful in determining source regions of precursor pollutants causing ozone violations. This type of transport analysis is necessary in order to understand the ozone problems in New England. Indeed the Airshed and ROM models can provide answers on the

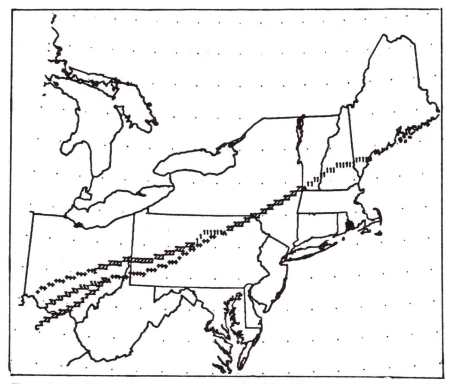

Figure 6. Backward trajectory output from the Branching Atmospheric Trajectory (BAT) model. Trajectory terminates at Cape Elizabeth, Maine on June 11, 1984 at 1500 GMT. Trajectory originated in the Ohio Valley 60 hours earlier.

transport question, but BAT and ATAD are far more economical to execute than Airshed or ROM. A typical BAT or ATAD computer run is less than $50. Of course neither BAT nor ATAD predict ozone, because neither has an ozone chemistry module; nevertheless, they can be an aid in resolving ozone transport issues. Unfortunately, there appears to be a bias in the results from both ATAD and BAT on ozone-conducive days in New England that may make their use for ozone analysis difficult. This possible bias is discussed below.

Discussion of the Possible Bias in the BAT and ATAD Models

Both the BAT and ATAD models have generally shown that the source regions affecting southern Maine are on a line approximately from Maine through extreme western Massachusetts into Pennsylvania (Fig. 6, for example). The models rarely implicate the Boston or New York metropolitan areas as the source of Maine's ozone. This is in conflict with the sea breeze hypothesis discussed above.

A possible explanation of this discrepancy lies in the vertical wind interpolation scheme used in both the BAT and ATAD models.

As discussed above, the weather most conducive to ozone formation in New England is in the summer when the Bermuda high is well established. During such events the surface flow over New England is southwesterly, and there is warm air advection in the lower layers of the atmosphere. The dynamics of the atmosphere during warm air advection require the winds to veer with height; therefore, the winds aloft during Bermuda high events are westerly. The winds aloft are also stronger than the surface winds, which are slowed by friction. Since the ATAD and BAT model use a vertically averaged wind for calculating their backward trajectories, and that vertically averaged wind is unduly influenced by the strong westerly winds aloft, a backward trajectory during these conditions will show a westerly bias. This bias may explain why the BAT and ATAD models generally show extreme western Massachusetts (Fig. 6) as the source region for southern Maine's ozone exceedances, rather than eastern Massachusetts (Fig. 1).

More sophisticated trajectory models might be better able to define actual source regions for the exceedances in Maine; however, the lack of upper-air data, both temporally (data are only available every 12 hours) and spatially (data are measured at only two upper-air stations in southern New England), will always affect the accuracy of any meteorological model, including BAT, ATAD, and ROM. Surface wind data, which are available hourly, and at a far greater resolution than upper-air data, may help better resolve atmospheric trajectories, but there still exists a large data void over the ocean. Also, surface wind data alone may not be representative of the winds that transport ozone and its precursors.

SUMMARY

An analysis of monitored ozone levels has shown that the NAAQS for ozone is often exceeded in New England. All six states have measured exceedances of the standard at some time during the past seven years. Connecticut is by far the New England state with the highest ozone levels and the most ozone exceedance days. The three northern New England states tend to have far better air quality than the three southern New England states (with respect to ozone), and it appears that the occasional excursion over the ozone standard in the three northern states is probably caused by transport and is not locally generated.

The chemistry of ozone formation, in conjunction with monitored ozone from remote areas around the globe and nineteenth century monitoring data, strongly suggest that the causes of the exceedances of the NAAQS observed in the United States are anthropogenic emissions of the ozone precursors, NO_x and VOC. The only way to achieve attainment with the standard is to reduce emissions of these precursors. Ozone chemistry, however, tells us that NO_x controls are not the most prudent method for lowering ambient ozone, but VOC controls are. The nonlinearity of ozone chemistry complicates the problem of determining the quantity of VOC reduction necessary to achieve attainment. However, through ozone modeling, especially with sophisticated regional ozone models such as ROM and Airshed, the difficult questions as to where and by how much VOC emissions should be reduced in order to achieve attainment with the NAAQS, can be answered.

This chapter also discusses how meteorology and ozone exceedance days are correlated. Ozone exceedance days show a strong positive biserial correlation (r_{bis}) to maximum temperatures ($r_{bis} = 0.78$). Exceedance days are only weakly correlated to percentage of possible sunshine ($r_{bis} = 0.28$), and show only a very weak inverse biserial correlation to mean wind speed ($r_{bis} = -0.16$). These results are in agreement with previous studies correlating ozone levels and meteorology. The correlations also make physical sense.

The correlation between meteorology and ozone is important in analyzing New England ozone trends. An analysis of New England ozone levels since 1980 shows the existence of a significant decrease in the mean second-maximum ozone concentration between 1983 (165 ppb) and 1986 (124 ppb). VOC emissions declined during this same period, leading some to speculate that the VOC emission reductions are responsible for the decline in ozone levels. Meteorological data show that two summers (1985 and 1986) were colder than normal and that the summer of 1983 was much warmer than normal in New England. Since high ozone levels are positively correlated with maximum temperatures, the recent downward trend in ozone is probably due to the cool summer weather, and precursor emission reductions play only a subordinate role.

Recommendations

The need for continued ozone research is clear. A better understanding of the meteorology conducive to ozone formation and transport will aid in regional ozone model development. More sophisticated and better controlled smog chamber studies will allow for the addition of better chemistry algorithms in regional ozone models. The need for accurate emission inventories for input to regional ozone models is essential. More meteorological data, both temporally and spatially, would be helpful in regional ozone models.

The need for continued ozone monitoring is also clear. A reliable long-term ozone trend network must be maintained. Additional ozone monitoring sites and meteorological monitoring sites should be positioned along the Atlantic Coast from Connecticut to Maine in order to track the ozone plumes from the large metropolitan areas located on the coastline. Finally, if the ROM or Airshed model are run for New England, the results should be used to locate additional monitors in the areas where the model predicts the highest concentrations, both on a one-hour and possibly an eight-hour, averaging basis.

ACKNOWLEDGMENTS

The author would like to thank Stephen Perkins for his help and guidance throughout this work. Sincere thanks are due to Allen Oi and Cheryl O'Halloran for helping with the data retrievals; to Steven Belcyzk for preparing the figures; and to Michael McKittrick, Heidi Maggelet, and Joan Fielding for typing the report.

DISCLAIMER

This chapter represents the views of the author and does not necessarily represent the official view of the U.S. Environmental Protection Agency. The mention of trade names or commercial products does not constitute endorsement or recommendation for use.

REFERENCES

1. "National Air Quality and Emissions Trends Report, 1984," EPA-450/4-86-001 (1986).
2. Yarn, J., W. Beal, and C. Tate, "Maps Depicting Nonattainment Areas Pursuant to Section 107 of the Clean Air Act—1985," EPA-450/2-85-006 (1985).
3. "Ozone in the Lower Atmosphere: A Threat to Health and Welfare," EPA Report-OPA-86-10 (1986).
4. Tingey, D. T. "Ozone Induced Alterations in Plant Growth and Metabolism," International Conference on Photochemical Oxidant Pollution and Its Control Proceedings: Volume II, EPA-600/3-77-001b, pp. 601–609 (1977).
5. Greene, C. Personal communication (1987).
6. Logan, J. A. "Tropospheric Ozone: Seasonal Behavior, Trends, and Anthropogenic Influences," *J. Geophys. Res.* 90(D6):10,463–10,482 (1985).
7. Dodge, M. C. "Effects of Selected Parameters on Predictions of a Photochemical Model," EPA-600/3-77-48 (1977).
8. Wark, K., and C. F. Warner. *Air Pollution: Its Origin and Control* (New York: IEP, 1976), pp. 395–416.
9. "Ozone and Other Photochemical Oxidants," EPA-600/1-76-027a, Chapter 2, (1976).
10. Altshuller, A. P. "The Role of Nitrogen Oxides in Nonurban Ozone Formation in the Planetary Boundary Layer Over N. America, W. Europe and Adjacent Areas of Ocean," *Atmos. Envir.* 20(2):245–268 (1986).
11. Alley, F. C., and Ripperton, L. A., "The Effects of Temperature on Photochemical Oxidant Production in a Bench Scale Reaction System," *JAPCA* 11(12):581–584 (1961).
12. Kwetz, B. A., S. Dennis, R. Donaldson, H.M. O'Brien, B.S. Porter, D. Ernst, and J. Dalton, "Massachusetts 1982 State Implementation Plan for Ozone and Carbon Monoxide," Massachusetts Department of Environmental Quality Engineering, August, 1982.

13. Rhoads, R. G. "Recommendations Concerning O_3 Program Design Criteria," USEPA Memorandum, Office of Air Quality Planning and Standards, Research Triangle Park, October 1, 1986, p. 27.
14. Dodge, M. C. "Chemistry of Oxidant Formation. Implication for Designing Effective Control Strategies," presented at North American Oxidant Symposium, Quebec City, February 25–27, 1987.
15. Spicer, C. W., J. L. Gemma, and P. R. Stickel, "The Transport of Oxidant Beyond Urban Areas. Data Analysis and Predictive Models for the Southern New England Study," EPA-600/3-77-041 (1977).
16. Spicer, C. W., D. W. Joseph, P. R. Stickel, and G. F. Ward, "Ozone Sources and Transport in the Northeastern United States," *Environ. Sci. Technol.* 13(8):975–985 (1979).
17. Siple, G. W., C. K. Fitzsimmons, K. F. Zeller, and R. B. Evans, "Long Range Airborne Measurements of Ozone Off the Coast of the Northeastern United States," *International Conference on Photochemical Oxidant Pollution and Its Control. Proceedings: Volume I*, EPA-600/3-77-001a, pp. 249–258 (1977).
18. "Northeast Corridor Regional Modeling Project. Aircraft Measurements. Boston and Vicinity," EPA-450/4-81-013 (1981).
19. Perkins, S. S. Personal communications (1987).
20. Westberg, H., E. Robinson, D. Elias, and K. Allwine, "Studies of Oxidant Transport Beyond Urban Areas. New England Sea Breeze-1975," EPA-600/3-77-055 (1977).
21. Lyons, W. A., and H. S. Cole, "Photochemical Oxidant Transport: Mesoscale Lake Breeze and Synoptic-Scale Aspect," *J. Appl. Meteorol.* 15(7):733–743 (1976).
22. Reiter, E. R. "The Role of Stratospheric Import on Tropospheric Ozone Concentrations," in *International Conference on Photochemical Oxidant Pollution and Its Control. Proceedings: Volume I*, EPA-600/3-77-001a, pp. 393–410 (1977).
23. Singh, H. B., F. L. Ludwig, and W. B. Johnson, "Tropospheric Ozone: Concentrations and Variabilities in Clean Remote Atmospheres," *Atmos. Environ.* 12, 2185–2196 (1978).
24. Lamb, R. G. "A Case Study of Stratospheric Ozone Affecting Ground-Level Oxidant Concentrations," *J. Appl. Meteorol.* 16(8):780–794 (1977).
25. Viezee, W., W. B. Johnson, and H. B. Singh, "Stratospheric Ozone in the Lower Troposphere-II. Assessment of Downward Flux and Ground Level Impact," *Atmos. Environ.* 17(10):1979–1993 (1983).
26. Ludwig, F. L., and R. Maughan, "Atmospheric Processes Affecting Ozone Concentrations in Northern New England," SRI Report #6908, EPA-901/9-28-001 (1978).
27. Trainer, M., "Impacts of Natural Hydrocarbons in Rural Ozone: Modeling and Observations," presented at North American Oxidant Symposium, Quebec City, February 25–27, 1987.
28. Linvill, D. E., W. J. Holler, and B. Olson, "Ozone in Michigan's Environment 1876–1880," *MWR* 108(11): 1883–1891 (1980).
29. Bojkov, R. D., "Surface Ozone During the Second Half of the Nineteenth Century," *J. Clim. and Appl. Meteor.* 25(3):343–352 (1986).
30. Wolff, G. T., P. J. Lioy, and R. S. Taylor, "The Diurnal Variation of Ozone at Different Altitudes on a Rural Mountain in the Eastern United States," *JAPCA* 37(1):45–48 (1987).
31. Bruntz, S. M., W. S. Cleveland, B. Kleiner, and J. L. Warner, "The Dependence of Ambient Ozone on Solar Radiation, Wind, Temperature, and Mixing Height,"

Preprint: Symposium on Atmospheric Diffusion and Air Pollution (Boston: American Meteorological Society), (1974), pp. 125–128.

32. Horie, Y., J. Cassmassi, L. Lai, and L. Gurtowski, "Weekend/Weekday Differences in Oxidants and Their Precursors," EPA-450/4-79-013 (1979).

33. Karl, T. R., and G. A. DeMarris, "Meteorological Conditions Conducive to High Levels of Ozone," in *International Conference on Photochemical Oxidant Pollution and Its Control. Proceedings: Volume I* EPA-600/3-77-001a, (1979), pp. 75–88.

34. "Procedures for Developing an Index of Meteorological Conditions Conducive to Exceedances of the Ozone NAAQS," USEPA Monitoring and Data Analysis Division, Draft Report, April, 1985.

35. Local Climatological Data, Monthly Summary, Hartford, Connecticut, from National Climatic Data Center, Asheville, NC.

36. Panofsky, H. A., and G. W. Brier, "Some Applications of Statistics to Meteorology," (University Park: Pennsylvania State University, 1968), pp. 98–100.

37. Whitten, G. Z., and H. Hogo, "User's Manual for Kinetic Model and Ozone Isopleth Plotting Package," EPA-600/8-78-014a (1978).

38. Reynolds, S. D., T. W. Tesche, and L. E. Reid, "An Introduction to the SAI Airshed Model and Its Usage," Systems Applications, Inc., Report EF78-53R4-EF79-31 (1979).

39. "Application of the Urban Airshed Model to the New York Metropolitan Area," New York State Department of Environmental Conservation, Draft Report #CX-811945 (1987).

40. Pearson, J. L. Personal communications (1987).

41. Lamb, R. G. "Numerical Simulations of Photochemical Air Pollution in the Northeastern United States: ROM1 Applications," USEPA-600/3-86/038 (1986).

42. Heffter, J. L. "Air Resources Laboratory Atmospheric Transport and Dispersion Model (ARL-ATAD)," NOAA Tech. Memorandum ERL ARL-81 (1980).

43. Heffter, J. L., "Branching Atmospheric Trajectory (BAT) Model" NOAA Tech. Memorandum ERL ARL-121 (1983).

44. Ludwig, F. L. "Ozone in Maine and Its Origins," SRI International, Report 7989 (1979).

45. Gordon, N. Personal communications (1987).

CHAPTER 4

Effects of Ozone on Crops in New England

William J. Manning

INTRODUCTION

Ozone (O_3) is the most important phytotoxic air pollutant in New England. It is a normal component of the troposphere, at low concentrations between 0.02 and 0.04 ppm. When ambient O_3 concentrations exceed this range, however, the potential for vegetation injury increases. This happens frequently during the spring, summer, and autumn in Massachusetts and other parts of New England.[1]

Low concentrations of O_3 result when the NO_2 dissociates to form nitric oxide (NO) molecules and molecular oxygen (O). Molecular oxygen atoms combine with O_2 molecules to form O_3. O_3 is unstable and reacts with NO to form NO_2 again. In this process, little build-up of O_3 occurs. When anthropogenic non-methane hydrocarbons, from combustion of fossil fuels and automobile emissions, enter the photolytic cycle, O_3 can build up. Hydrocarbons react with NO to form an additional secondary photochemical oxidant peroxyacetyl nitrate (PAN). With little NO to interact with, O_3 formation continues, and elevated concentrations of O_3 occur.[2,3]

Due to the phenomenon of long-range transport[3,4,5,6] O_3 and/or its precursors can move very long distances, resulting in regional, rather than localized, air quality problems. O_3 concentrations in remote rural areas can be much higher than in or near cities and industrial areas.[3,4]

The Environmental Protection Agency (EPA) requires states to continuously monitor air to determine compliance with the Federal Clean Air Act. In New England, all of Massachusetts, Connecticut, and Rhode Island, and large areas of southern New Hampshire and Maine are in violation of the National Ambient Air Quality Standard for O_3. The standard (0.12 ppm hourly average for one hour) is violated on numerous occasions in Massachusetts.[1] Verified and possible effects of these elevated concentrations of O_3 on crops in New England will be discussed below.

EFFECTS OF OZONE ON CROP PLANTS

Symptoms of Ozone Injury

O_3 enters plant leaves via open stomates during the normal process of gas exchange between a leaf and its immediate environment.[7] Palisade parenchyma cells may be injured first, with spongy mesophyll cells affected later. Injury is often more severe on upper leaf surfaces, but may also occur on lower surfaces. Leaves that are still expanding, or have just achieved full size, are the most susceptible to O_3. For this reason, O_3 injury is usually found on older leaves, and not on young leaves.[2,8] O_3 may cause acute injury, chronic injury, or it may affect growth and yield, with or without visual symptoms.[9] Acute injury occurs on plants exposed to high concentrations of O_3 for a short time period. Chronic injury results from long-term exposure of plants to low concentrations of O_3. In nature, both types of symptoms may occur on the same plant, due to the fluctuating nature of ambient concentrations of O_3.

Depending on the type and variety of plant, the concentration and duration of O_3 can cause a wide array of symptoms on plants. These have been summarized in several reviews and texts.[2,8,10-13] Some common symptoms observed on crop plants in Massachusetts are summarized in Table 1.

During the growing season, certain varieties of common crop plants may exhibit typical symptoms of O_3 injury in Massachusetts. These are summarized in Table 2.

Table 1.   Common Symptoms of O_3 Injury on Leaves of Crop Plants.

Acute Injury	Chronic Injury
Flecking: Small necrotic areas due to death of palisade cells, metallic or brown, fading to tan, gray or white	*Pigmentation (bronzing):* leaves turn red-brown to brown as phenolic pigments accumulate
Stippling: tiny punctate spots where a few palidade cells are dead or injured, may be white, black, red or red-purple	*Chlorosis:* may result from pigmentation or may occur alone as chlorophyll breaks down
	Premature Senescence: early loss of leaves or fruit

Table 2. Crop Plants Commonly Affected by O_3 and Typical Symptoms Expressed.

Plant	Foliar Symptoms
Beans (*Phaseolus*)	Bronzing and chlorosis
Cucumber (*Cucumis*)	White stipple
Grape (*Vitis*)	Red to black stipple
Morning glory (*Ipomoea*)	Chlorosis
Onion (*Allium*)	White flecks and tips
Potato (*Solanum*)	Grey fleck and chlorosis
Soybean (*Glycine*)	Red-bronze and chlorosis
Spinach (*Spinacea*)	Grey fleck to white
Tobacco (*Nicotiana*)	Metallic fleck to white
Watermelon (*Citrullus*)	Grey fleck

Factors Affecting Ozone Injury on Crops

Crop plants are exposed to fluctuating concentrations of O_3 and environmental conditions throughout their life cycles. This means that there will be periods when O_3 concentrations are too low for sensitive plants to respond, and periods where environmental conditions prevent or reduce O_3 uptake. During these periods, compensatory growth may occur, which may eliminate any effects of O_3 at harvest in terms of yield effects.

Warm temperatures, sunlight, high relative humidity, good nutrition and adequate soil moisture, and other factors are necessary for O_3 injury to occur.[10] Sensitive plants that are not suffering from other stresses are usually injured by O_3.

Heagle, et al.[14] and others have demonstrated that plants often have different thresholds for foliar O_3 injury and yield losses. Threshold concentrations of O_3 that caused foliar injury on field corn hybrids were different than thresholds for yield losses. Reich and Amundsen[15] report yield losses due to O_3 in the absence of foliar symptoms. The correlation between foliar injury symptoms and yields is usually not very good.

Secondary Effects of Ozone on Plants

In addition to causing visible symptoms on leaves and reducing the growth and yield of aboveground plant parts, O_3 can have secondary effects on root development and predisposition of plants to insects and diseases. These must be considered when determining the effects of O_3 on crops.

Chronic O_3 injury results in less translocation of photosynthate from shoots to roots. This may affect yields, due to reduced water and mineral uptake and enhanced senescence of roots.[16]

Acute O_3 injury on potato and geranium leaves can be invaded by the common necrotrophic pathogenic fungus *Botrytis cinerea*, allowing extensive fungal foliar blight to occur, where it would not have been possible before.[17,18] It would also obscure the role of O_3 in disease initiation and effects on crop yield.

Examples of Ozone Injury on Crop Plants

In the 1960s and early 1970s, there was considerable research to determine the effects of O_3 on crops in Massachusetts and Connecticut, but not in the other New England states. A regional survey for O_3 injury in the New England region was conducted for one growing season.[19]

Most investigations have centered on describing O_3 effects on plants, determining possible effects on yields, and investigating possible interactions with plant pathogenic fungi.[17,18,20-22] A few examples of the verified effects of O_3 on crop plants in Massachusetts are given here.

Feder and Campbell[23] grew carnations (*Dianthus caryophyllus* L. 'White Sim') in greenhouse chambers and exposed them to varying doses of O_3. When plants in the flower initiation stage were exposed to 0.075 ppm O_3 for 10 days, no foliar symptoms occurred, but bud formation was depressed.

In certain parts of Massachusetts, and in northern New Jersey, spinach (*Spinacea oleraceae*) is no longer a viable crop. It became too difficult to market plants with extensive white flecking and chlorosis on older leaves. O_3 was shown to be the cause of the problem in greenhouse tests with six spinach cultivars. Two of the cultivars were less sensitive to O_3 and could be recommended to replace the original cultivar.[20]

Damicone, et al.[24] demonstrated that soybean [*Glycine* max (L.) Merr.] genotypes of maturity groups 00, 0, and I responded in predictable ways to ambient O_3. Evaluation of progeny from reciprocal crosses between O_3-sensitive and O_3-resistant lines, showed that O_3 tolerance was inherited qualitatively. This was the first report demonstrating heritability of foliar O_3 injury tolerance in soybeans.

Suppression of foliar O_3 injury on field-grown snap beans (*Phaseolus vulgaris* L.), by weekly applications of the fungicide benomyl, allowed determination of O_3-caused yield losses, under natural field conditions, without artifacts from exposure chambers.[21] Results were compared for the O_3-susceptible cultivar 'Tempo' and the O_3-tolerant cultivar 'Slenderwhite' (Table 3). Ozone concentrations were high enough and persistent enough (335 hourly average readings of 0.04 ppm and higher) throughout the season to cause yellowing, bronzing and premature senescence of Tempo leaves. Foliar sprays of benomyl reduced symptoms and premature senescence of Tempo leaves by 75 to 80%. Nonsprayed plants of both cultivars bloomed earlier and bore pods earlier than sprayed plants. This is reflected in yield figures for week 1 (Table 3). Significantly greater numbers and weights of pods were obtained from sprayed Tempo plants at harvests 2 and 5 and for total yield. Significant differences were not observed between any average yields from sprayed and nonsprayed plants of Tenderwhite.

Table 3.   Influence of Foliar Applications of Benomyl on Growth and Yield of Snap Beans.

Cultivars and Treatments	Average Yields by Weeks[d]					Average Total Yields
	1	2	3	4	5	
Tempo[a]						
No. of pods:						
sprayed[b]	77.3	213.2*	504.3	326.3	226.8**	1347.9**
non-sprayed	91.0	151.7*	417.3	234.0	127.3**	1021.3**
Wt. of pods (g):						
sprayed	180.6	712.3*	1589.7	1225.1	1035.8*	4743.5**
non-sprayed	226.0	433.0*	1457.2	831.8	411.0*	3359.0**
Tenderwhite[c]						
No. of pods:						
sprayed	96.8	335.7	264.7	373.5	181.5	1252.2
non-sprayed	115.0	330.0	249.7	330.7	116.3	1141.7
Wt. of pods (g):						
sprayed	354.3	1166.0	1011.6	1700.0	647.6	4879.5
non-sprayed	428.3	1529.5	977.6	1435.1	465.1	4835.6

[a]First harvest made 54 days after planting.
[b]Weekly foliar sprays of benomyl at 2.4 g1^{-1}.
[c]First harvest made 61 days after planting.
[d]Average per four replications, twenty plants each.
*Significantly different values (P = 0.05), standard analysis of variance.
**Significantly different values (P = 0.01), standard analysis of variance.

Cooley and Manning[25] applied the techniques of plant growth analysis to determine the effects of long-term exposure of alfalfa (*Medicago Sativa* L.) cultivars to low concentrations of O_3. Ozone reduced photosynthate production, and increased the proportion of photosynthate apportioned to stems and leaves, at the expense of roots and crowns. This could lead to increased winter injury and invasion of roots and crowns by pathogenic fungi.

These few examples illustrate how a variety of methods have been used to determine the effects of O_3 on crop plants in Massachusetts.

The NCLAN Network

This presentation would not be complete without some mention of the National Crop Loss Assessment Network (NCLAN). Although it did not operate in New England, results from Ithaca, New York and other locations on the eastern seaboard are often related to New England.

The EPA initiated NCLAN in 1980 to form a nationwide network to determine pollutant (primarily O_3) effects on crop plants. Large numbers of cylindrical open-top chambers are placed over growing crops in the field. Air movement systems allow exclusion or inclusion of pollutants at will for the life cycle of the crop. After a 6-year period, significant yield losses due to O_3 are claimed for soybeans, corn, wheat, and cotton. The scale of the network, the methods used to obtain the data, and the evaluation of the results, however, are not universally accepted and have been criticized.[26]

OUTLOOK FOR NEW ENGLAND

Until regional air quality strategies are employed, it is not likely that most of New England's concentrations of O_3 will decrease. In the meantime, few studies on the effects of O_3 on crops in the region will be conducted. What money that is available will be spent on research on O_3 effects on trees. If Reich and Amundson[15] are correct in their conclusions that O_3-caused reductions in net photosynthesis are occurring over much of the eastern United States, and that fast-growing plants, like most crops, have high O_3 uptake rates, we can expect additional declines in crop plant yields in New England.

REFERENCES

1. Furfey, R. N., and J. A. Doucett. "Ozone Concentrations in Massachusetts: Potential Impact on Vegetation," Massachusetts Dept. of Environmental Quality Engineering, Division of Air Quality Control (1985).
2. Lacasse, N. L., and M. Treshow. *Diagnosing Vegetation Injury Caused by Air Pollution,* (U.S. Environmental Protection Agency Handbook, 1976).
3. *Ozone and Other Photochemical Oxidants*, National Academy of Sciences (1977), 719 pp.
4. Cleveland, W. S., B. Kleiner, J. E. McRae, and J. L. Warner. "Photochemical Air Pollution: Transport from the New York City Area into Connecticut and Massachusetts," *Science*: 191:179–181 (1976).
5. Cleveland, W., and T. Graedel. "Photochemical Air Pollution in the Northeast United States," *Science* 204:1273–1278 (1979).
6. Lee, S. D. *Evaluation of the Scientific Basis for Ozone/Oxidants Standards* (Pittsburgh: Air Pollution Control Association, 1985).
7. Rich, S., P. E. Waggoner, and H. Tomlinson. "Ozone Uptake by Bean Leaves," *Science* 169:79–80 (1970).
8. Manning, W. J., and W. A. Feder. *Biomonitoring Air Pollutants with Plants* (London: Applied Science Pub., Ltd., 1980).
9. Unsworth, M. H., and D. P. Ormrod. *Effects of Gaseous Air Pollution in Agriculture and Horiculture* (London: Butterworth Scientific, 1982).
10. Heck, W. W. "Factors Influencing Expression of Oxidant Damage to Plants," *Annual Review of Phytopathology* 6:165–188 (1968).
11. Hill, A. C., H. E. Heggestad, and S. N. Linzon. *Recognition of Air Pollution Injury to Vegetation: A Pictorial Atlas* (Pittsburgh: Air Pollution Control Association, 1970).

12. Mansfield, T. A. *Effects of Air Pollutants on Plants,* (Cambridge University Press, 1976).
13. Mudd, J. B., and T. T. Kozlowski. *Responses of Plants to Air Pollution,* (New York: Academic Press, 1975).
14. Heagle, A. S., R. B. Philbeck, and W. M. Knott. "Thresholds for Injury, Growth, and Yield Loss Caused by Ozone on Field Corn Hybrids," *Phytopathology* 69:21–26 (1979).
15. Reich, P. B., and R. G. Amundson. "Ambient Levels of Ozone Reduce Net Photosynthesis in Tree and Crop Species." *Science* 230:566–570 (1985).
16. Manning, W. J. "Chronic Foliar Ozone Injury: Effects on Plant Root Development and Possible Consequences," *California Air Environment* 7:3–4 (1978).
17. Manning, W. J., W. A. Feder, I. Perkins, and M. Glickman." Ozone Injury and Infection of Potato Leaves by *Botrytis Cinerea,"* *Plant Disease Reporter* 53:412–415 (1969).
18. Manning, W. J., W. A. Feder, and I. Perkins. "Ozone Injury Increases Infection of Geranium Leaves by *Botrytis cinerea,"* *Phytopathology* 60:669–670 (1970).
19. Naegele, J. A., W. A. Feder, and C. J. Brandt. "Assessment of Air Pollution Damage to Vegetation in New England," University of Massachusetts (1972).
20. Manning, W. J., W. A. Feder, and I. Perkins. "Sensitivity of Spinach Cultivars to Ozone," *Plant Disease Reporter* 56:832–833 (1972).
21. Manning, W. J., W. A. Feder, and P. M. Vardaro. "Suppression of Oxidant Injury by Benomyl: Effects on Yields of Bean Cultivars in the Field," *J. Environ Quality* 3:1–3 (1974).
22. Rich, S., and A. Hawkins. "The Susceptibility of Potato Varieties to Ozone in the Field," *Phytopathology* 60:1309 (1970).
23. Feder, W. A. and F. J. Campbell. "Influence of Low Levels of Ozone on Flowering of Carnations," *Phytopathology* 58:1038–1039.
24. Damicone, J. P., W. J. Manning, S. J. Herbert, and W. A. Feder. "Foliar Sensitivity of Soybeans from Early Maturity Groups to Ozone and Inheritance of Injury Response," *Plant Disease* 71:332–336 (1987).
25. Cooley, D. R., and W. J. Manning. "Effects of Ozone on the Dynamics of Growth and Assimilate Partitioning in Alfalfa, *"Proc. 7th World Clean Air Congress, Sydney Australia*, Vol. 3, pp. 140–147.
26. Krupa, S., and R. N. Kickert. "An Analysis of Numerical Models of Air Pollutant Exposure and Vegetation Response," *Environmental Pollution* 44 (1987).

CHAPTER 5

Effects of Ozone on Forests in the Northeastern United States

Mary Beth Adams and George Evans Taylor, Jr.

INTRODUCTION

Atmospheric pollutants emitted from localized sources have been recognized for decades as an influence on forest ecosystems in eastern North America, but only recently has research begun to focus on the long-term responses of forests to regionally distributed air pollution stress. Of particular concern to this conference is the role of ozone (O_3) and other photochemical pollutants. While the emphasis is O_3, it is recognized that other chemical pollutants are deposited to forest canopies through a combination of wet and dry deposition. Because many of these other pollutants exhibit the same spatial distribution across the region as O_3, it is difficult to accurately attribute any regional forest response to a single pollutant. Accordingly, to address the specific role of O_3 in the Northeast, we have focused on characterizing (a) the unique properties of O_3 as an ecological stress in terrestrial ecosystems, (b) the region's O_3 air quality from an ecophysiological perspective, (c) physiological and growth responses of individual trees to O_3, and (d) forest stand responses to O_3.

CHARACTERISTICS OF OZONE AS AN ECOLOGICAL STRESS

As an environmental stress in forested landscapes, O_3 is distinct from other stresses of anthropogenic origin. Following deposition, O_3 molecules are ephemeral, reacting quickly and decomposing to oxygen and free radicals. In contrast to the biogeochemical behavior of residuals of sulfur and nitrogen oxides, fluorides, heavy metals, and radionuclides, O_3 by-products are not selectively retained or accumulated in particular components of the ecosystem. Whereas the half-life of many anthropogenic pollutants in natural ecosystems is typically measured in timeframes approaching years, decades, or even centuries, the half-life of O_3 is on the order of minutes. Moreover, because O_3 phytotoxicity is a consequence of the production of free radicals,[1] biological systems are not challenged with a completely novel chemical stress since free radicals are a common byproduct of aerobic metabolism. Finally, the distribution of O_3 is truly regional (Figure 1) so that ecological and physiological effects are not confined to areas proximal to major point sources such as occurs for emissions of sulfur oxides or heavy metals.

The unique ecological and toxicological characteristics of O_3 present a significant problem in evaluating the pollutant's effects on natural ecosystems because the conceptual framework and experimental approaches to addressing the response of terrestrial ecosystems to similar ecological stresses evolved from studies in which the level of stress was either acute or very localized. The notable examples are from areas experiencing ionizing radiation, smelter emissions, or point sources of sulfur oxides. In these acutely stressed areas, the pattern of ecosystem response is broadly predictable and is summarized as a simplification of community structure according to the species-specific ratio of carbon assimilation to respiration and successional dynamics.[2] While this framework is a valuable reference in understanding ecosystem-level responses to acute stresses, it may not be the most appropriate basis from which to evaluate long-term ecosystem responses to nonaccumulating, intermittent, chronic-level stresses that are regionally distributed, all characteristics of O_3.

OZONE CLIMATOLOGY FROM
AN ECOPHYSIOLOGICAL PERSPECTIVE

Although interest in characterizing O_3 climatology has increased, the temporal (e.g., diurnal, monthly, seasonal, annual) and spatial (e.g., regional, elevational) attributes of O_3 in natural ecosystems of North America are not fully resolved. Even though existing databases may prove useful in addressing some issues of O_3 climatology,[3] the geographical distribution of monitoring sites does not necessarily coincide with many of the forested landscapes of greatest ecological concern. An equally important constraint underlying the dearth of ecologically relevant analyses is the assumption that methodologies developed in the disciplines of atmospheric science or human health are legitimate bases to characterize O_3 climatology in nonurban, remote locations. We propose that many of these methods cannot adequately characterize the dynamics of O_3 air quality in ways that are pertinent to understanding the pollutant's ecological and physiological effects.

ORNL–DWG 87-1737

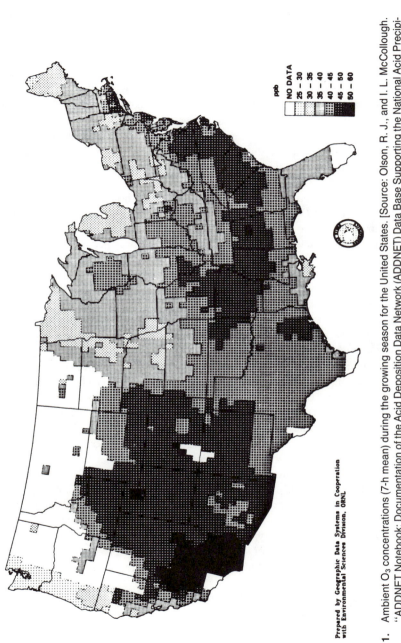

Prepared by Geographic Data Systems in Cooperation
with Environmental Sciences Division, ORNL

Figure 1. Ambient O_3 concentrations (7-h mean) during the growing season for the United States. [Source: Olson, R. J., and I. L. McCollough. "ADDNET Notebook: Documentation of the Acid Deposition Data Network (ADDNET) Data Base Supporting the National Acid Precipitation Assessment Program," ORNL/TM-10086, Oak Ridge National Laboratory, Oak Ridge, Tennessee.]

Statistical methodologies for characterizing O_3 in forest ecosystems must recognize the (1) temporal and spatial characteristics of O_3 that are unique from those in urban areas, (2) ecological and physiological attributes that influence O_3 effects on individual trees and forest communities, and (3) unique chemical attributes of O_3 that necessitate analyses distinct from those used for other pollutants (e.g., SO_2, NO_2). Although the first concern has been discussed,[3-5] the significance of the second and third is less appreciated in current efforts to assess the influence of O_3 on natural ecosystems.

A physiologically based conceptual model of plant response[1] to O_3 provides a framework for identifying those features of O_3 climatology that are physiologically relevant. The model identifies three critical processes governing the extent of plant injury due to O_3: (1) foliar uptake of O_3 molecules, (2) intrinsic sensitivity of critical metabolic processes, and (3) role of homeostatic mechanisms.

Foliar uptake of O_3 molecules is important because the physiological sites of injury are cells of the leaf interior, including those of the stomatal complex and photosynthetically active mesophyll tissues. Any species-specific or environmental factor that affects O_3 transport in the gas phase (e.g., stomatal conductance) or liquid phase (e.g., free radical scavenging) will influence the concentration of toxic O_3 derivatives at sensitive physiological sites. One of the principal factors governing O_3 uptake is stomatal conductance,[6] which exhibits characteristic diurnal patterns that are species-specific and superimposed on pronounced seasonal phenology. Over a 24-h period during the growing season, peak conductance rates are commonly observed in the mid-afternoon and extend into the early evening hours, while minima are in the late evening and early morning hours. For *Pinus taeda* grown under field conditions, stomatal conductance typically achieves 50% maximum before 0900 h and continues at or above that rate until 2000 h (Figure 2c). While this is a common diurnal pattern,[7] there are notable variations among and within species and pronounced differences induced by environmental variables such as light, temperature, vapor pressure deficit, and soil water availability.[8]

One of the most sensitive physiological processes is net CO_2 assimilation,[5-9] which also exhibits characteristic diurnal patterns (Figure 2c). During the growing season, rates of CO_2 assimilation tend to rise in the mid-morning, achieve maxima in the mid- to late afternoon, and subsequently decline in the early evening hours (Figure 2c). Seasonality and the associated phenological changes in physiological processes play a dominant role in governing the rate of CO_2 assimilation, and because the effect of season is highly site- and species-specific, it must be recognized as an important factor governing the selection of O_3 exposure statistics as discussed elsewhere.[4]

Homeostatic processes are endogenous functions that either compensate for injury or repair metabolic dysfunctions induced by chemical or physical stresses.[10] The effectiveness of homeostasis is a function of the magnitude of the stress event and the duration of the intervening respite. In general, the magnitude of injury (e.g., O_3-induced reduction in CO_2 assimilation rate) is proportional to the intensity and duration of the stress event but inversely proportional to the length of

ORNL–DWG 87-1732

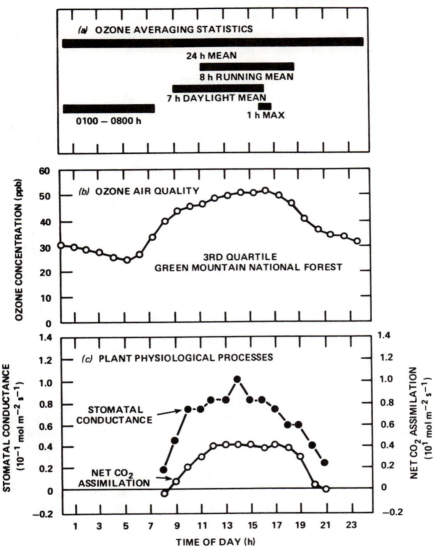

Figure 2. Diurnal relationships among ozone averaging statistics (a), pattern of ozone air quality in the Green Mountain National Forest (b), and key physiological processes in forest trees based on foliar data from *Pinus taeda* L. (c).

the respite period. Longer respite periods allow homeostatic processes to repair or compensate for injury through the initiation of new foliar tissues or the enhancement of CO_2 assimilatory capacity in uninjured tissues. Over time spans measured in hours, days, years, or decades, the frequency of stress events and the duration of respites are critical variables to understand how episodically elevated levels of O_3 influence the growth and development of forest trees.

Given these physiologically based criteria, any statistical analysis of O_3 air quality is more sound if it recognizes the (1) diurnal behavior in foliar physiological activity associated with stomatal conductance and net CO_2 assimilation, (2) seasonality of physiological and growth processes in terrestrial vegetation, and (3) importance of homeostatic processes in governing plant response.

Ozone concentrations in many natural ecosystems, particularly those in eastern North America, may equal or even exceed those in upwind urban airsheds within the same region, a phenomenon attributed to the pollutant's photochemical induction time, physical transport processes, additivity of polluted air masses, and reduced efficiency of O_3 scavenging processes in rural atmospheres.[4,8] Whereas peak O_3 levels tend to be greater in urban airsheds, many of the daily mean statistics are higher in more remote landscapes within the same region. The diurnal O_3 cycle in natural ecosystems may differ markedly from that of urban airsheds in which the highest concentrations are pronounced and routinely observed in the mid-afternoon. In remote areas, the maximum is often shifted into the late afternoon and evening hours or is altogether absent.[11,12] If elevated O_3 levels are shifted into the late afternoon and evening hours in remote locations, the appropriate statistic with which to characterize O_3 exposure dynamics may differ from the 7-h daylight mean used to evaluate crop loss. In addition, if a significant percentage of the peak exposure time is during periods of stomatal closure, the pivotal physiological role attributed to the pollutant's direct effects on carbon gain may warrant reappraisal, and alternative proposals for plant effects for which light is not a prerequisite (e.g., O_3 deposition, cuticular degradation, and enhanced rates of foliar leaching) may be ecologically significant.

The geographical distribution of O_3 concentrations during daylight hours of the growing season in the United States indicates substantial differences among regions of the country (Figure 1). The highest seasonal mean O_3 concentrations are in the Southwest and range from 50 to 60 ppb. Concentrations ranging from 45 to 50 ppb are common across a large segment of the arid West and along a band of the Southeast. In comparison, the O_3 concentrations in the Northeast tend to be lower, averaging ≤ 40 ppb for the 7-h daylight mean. These data suggest that O_3 effects on regional forest ecosystems are likely to be less in the Northeast than either the Southwest or Southeast. To obtain an example of the O_3 air quality in a forested landscape in northeastern North America, we analyzed the 1979–1981 data from the Green Mountain National Forest (GMNF). This site, located in Vermont at an elevation of 400 m above sea level, is in a rural area characterized by patches of forest and clearings. Because the dominant landscape component is forests, we refer to the area as a forested landscape. Our intent is to use this as a representative site in the Northeast to demonstrate how O_3 air quality can be analyzed from ecophysiological perspectives. For particular statistical analyses, the patterns of O_3 exposures from GMNF are compared with those from a nearby large urban airshed (Providence, Rhode Island) having comparable periods of reliable O_3 measurements. All analyses used O_3 concentration units of parts per billion (10^9 v/v).

Daily and seasonal changes in O_3 concentrations and corresponding patterns of critical physiological processes should be the basis for selecting the O_3 exposure statistics (Figure 2a and b). Based on the third quartile data (averaged over 1979–1981), O_3 concentrations are lowest in the predawn hours, increasing at 0700 h to a maximum of 1600 h, with falling concentrations thereafter until 2400 h (Figure 2b). The most commonly used statistics for characterizing O_3 exposures are means averaged over fixed time intervals of 0–2400 h (24-h mean), 0100–0800 h (predawn mean), and 1000–1700 h (7-h daylight mean) (Figure 2a). Other statistics, not dictated by a specified window of time, are the maximum 8-h running mean and the 1-h maximum (Figure 2a). The 7-h daylight mean was developed to characterize O_3 exposures for agricultural crop loss assessments, while the 8-h running mean has merit because it captures some of the aspects of the episodal periods of O_3 exposure. For this region-wide analysis of O_3 climatology in field situations, we have not used the concept of dose (i.e., product of concentration and time in ppb.h). This type of analysis has merit for addressing physiological effects under defined exposure conditions (e.g., square wave exposure profiles) but does not adequately characterize the dynamics of elevated O_3 concentrations in natural ecosystems.

Although the diurnal patterns in O_3 air quality in forested and urban landscapes in the same region of the Northeast have some common features, there are notable differences (Figure 3a and b). In the second quartile, mean O_3 concentrations in the GMNF consistently exceeded those in Providence by at least 20%, with even higher differences in the late afternoon and evening hours. Whereas peak O_3 concentrations in the urban airshed occurred at 1400 h, the maximum in the forested landscape was delayed until 1700 h. The third quartile exhibited a different pattern between the two locations (Figure 3b): in Providence, O_3 concentrations were consistently higher than those in the forest landscape except in the predawn hours, and the differences were most apparent in the afternoon, where the urban concentrations exceeded those in the forest by greater than 30%. As in the second quartile, peak O_3 exposure in the forest tended to be broad-based in comparison to the more accentuated single peak characteristic of the urban area. While the mean concentrations were higher in the third than in the second quartile in the urban landscape, they were lower in the more remote forested ecosystem.

The seasonality of O_3 exposure statistics in the GMNF is notable (Figure 4a, b, and c). The 24-h means were at a minimum (20–25 ppb) in the late fall and winter, and increased slowly thereafter to a late spring maximum in May at 40 ppb (Figure 4a). All subsequent monthly averages during the remainder of the growing season declined. The seasonality of the 7-h daylight mean was similar, except for the maximum being delayed one month to June (Figure 4b). On the average, the 7-h daylight mean concentrations during the growing season were

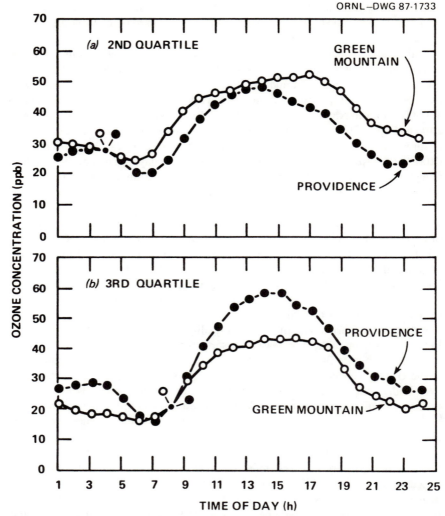

Figure 3. Diurnal pattern of O_3 concentrations at Green Mountain National Forest (rural landscape) and Providence, Rhode Island (urban landscape) for the second (a) and third (b) quartile.

more than 25% higher than the corresponding 24-h mean. The 8-h running mean (Figure 4c) exhibited even higher (10-15%) mean O_3 concentrations than the 7-h daylight mean, a consequence of the statistic being more responsive to O_3 episodes.

The frequency of sustained elevated O_3 levels (i.e., episodes) and the duration of intervening respite periods are important issues with respect to O_3 air quality,

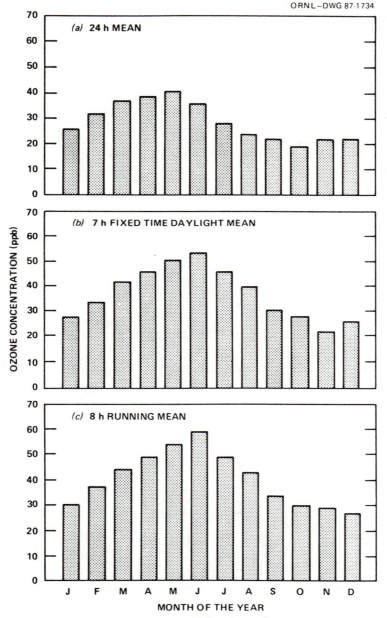

ORNL–DWG 87-1734

Figure 4.   Monthly statistics of O_3 air quality at Green Mountain National Forest using 24-h mean (a), 7-h fixed daylight mean from 1000–1700 h (b), and 8-h running mean (c).

given the physiological role for homeostasis. Based on the 1979–1981 data monitored at the GMNF, O_3 episodes during the growing season were characterized, defining an episode as any day in which a single 1-h O_3 concentration was ≥ 80 ppb. Consecutive days with such values were tallied as single events consisting of multiple days. The GMNF experienced an average of 12 episodes during the growing season, and the duration of individual events ranged from 1 to 5 consecutive days (Figure 5a). The probability of an O_3 episode changed during the

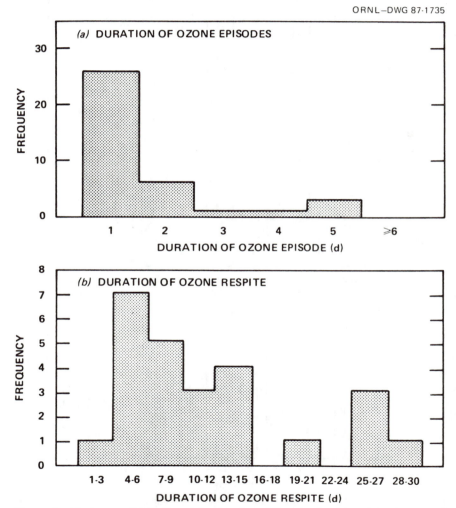

ORNL–DWG 87-1735

Figure 5. Frequency distribution of the duration of O_3 episodes (a) and the duration of O_3 respites (b).

course of the growing season, with events being most probable during the months of July, August, and September, and least probable early in the growing season (Figure 6a). The mean duration of O_3 episodes during the May to August period was ≥ 2 days, and the maximum (2.8 days) occurred in August (Figure 6b). In the GMNF, the months of July and August experienced the highest mean duration of O_3 events, combined with the highest probability of episodes. The combination of the two factors resulted in a specific pattern of days per month experiencing O_3 episodes, and approximately 40% of the days in July and August during the 1979–1981 period were classified as O_3 episodes (Figure 6c).

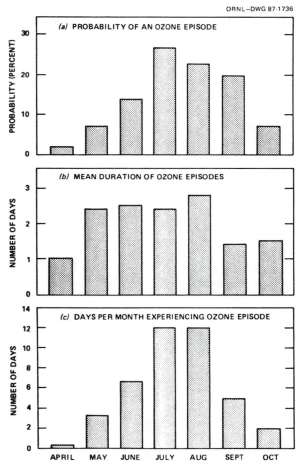

ORNL–DWG 87-1736

Figure 6. Growing season analysis of O_3 episodes at Green Mountain National Forest including probability of an episode (a), duration of episode (b), and days per month experiencing O_3 episodes (c).

Thus, while the 24- and 7-h mean O_3 concentrations indicate peak O_3 levels in the spring, episodically high O_3 levels are more common in the mid- to later segments of the growing season in this forested landscape in the Northeast.

Based on the three-year period, there is a 30% probability that any given episode during the growing season will last up to 2 days, whereas the probability of episodes being more than 2 days is ≤15% (Figure 7). Single day O_3 episodes are the most probable episodic event (P = 70%). The frequency distribution of respite periods indicates a median of 4 to 6 days in duration (Figure 5b). Given the frequency of respites, a second episode is highly probable within 10 days, and respites of less than 5 days are not uncommon (P = 25%). Thus, O_3 episodes in forested landscapes in the Northeast are separated temporally by respites that are far greater in duration than that of the O_3 episode. The characteristics of O_3 episodes in the urban Providence landscape are somewhat different, as evidenced by the higher probability of longer individual episodes and a greater probability that the respites will be longer in duration (Figure 7). Thus, the forested landscape in the same region has a higher probability of shorter periods of sustained elevated O_3 levels, but these episodes are separated by shorter respites. Based on the analysis of relevant physiological properties presented earlier, this characteristic of O_3 exposure would provide less opportunity for homeostatic processes to repair or compensate for injury in forest landscapes.

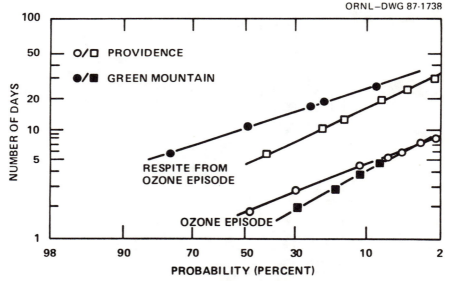

Figure 7.   Probability analysis of the duration of O_3 episodes and respites at a rural (Green Mountain National Forest) and urban (Providence, Rhode Island) landscape.

PHYSIOLOGICAL AND GROWTH RESPONSES OF INDIVIDUAL TREES TO OZONE

While the response of individual trees to O_3 stress is not synonymous with that of stands, communities, or ecosystems, an understanding of how O_3 affects the physiology and growth of individual trees is a requisite to addressing long-term changes in forest productivity. This section focuses on (a) the influence of O_3 on net photosynthesis, (b) the influence of O_3 on whole-plant carbon allocation to both aboveground (e.g., reproduction) and belowground (e.g., fine root growth) sinks, and (c) the empirical relationship between O_3 air quality and tree growth.

Influence of Ozone on Net Photosynthesis

Processes governing the assimilation of carbon dioxide are notably responsive to O_3 in a variety of coniferous and broadleaf tree species endogenous to the northeastern United States (Figure 8). The most distinctive feature of Figure 8 is the diversity of responses among and within species ranging from marked inhibition in sensitive species to a slight enhancement of photosynthesis in resistant species. Hybrid poplar (*Populus deltoides* X *trichocarpa*) appears to be sensitive, as relatively large decreases in net photosynthesis were recorded at low O_3 dosage. Conversely, red oak (*Quercus rubra* L.) and red spruce (*Picea rubens* Sarg.) appear to be O_3 resistant. In white pine (*Pinus strobus*) the response of photosynthesis ranges from negative to stimulatory effects, demonstrating the large intraspecific differences in O_3 resistance characteristic of tree species. This variability suggests that large intraspecific differences in O_3 resistance exist for some northeastern tree species.

Interspecific differences in O_3 effects on photosynthesis may be due to differences in intrinsic photosynthetic rates, gas-exchange characteristics, or physiological state.[5] Reich and Amundson[13] suggested that the greater susceptibility of hybrid poplar (*Populus tremuloides* X *trichocarpa*) may be due to the higher intrinsic rate of photosynthesis, a rate similar to that of some crop plants. An equally plausible explanation rests with the unusually high stomatal conductance of this species, which would enhance O_3 flux into the leaf interior. Boyer and coworkers[14] reported that white pine clones that were classified as "sensitive" to O_3 exhibited a higher photosynthetic rate (by approximately 36%) in clean air than "tolerant" clones. External mitigating factors such as moisture and nutrient status, season of the year, climate, and competition may also modify O_3 effects on net photosynthesis. Mild drought stress can serve to protect plants from O_3 damage,[15] while low vapor pressure deficit regimes enhance injury from air pollutants.[16,17] Also, sub- or supraoptimal nutrition will affect a plant's response to O_3.[9] As mentioned earlier, timing of and time between O_3 events is important because some species recover from O_3 damage more quickly than others.[4]

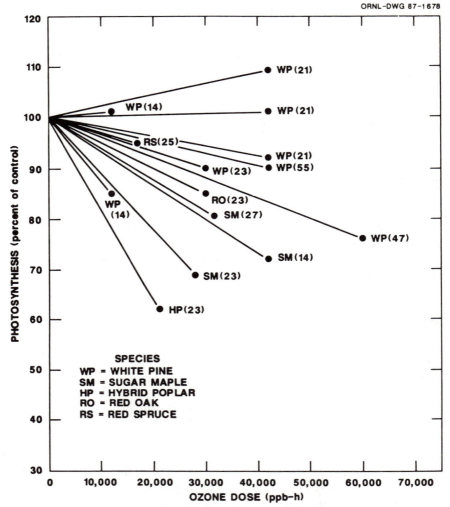

ORNL-DWG 87-1678

Figure 8. Effect of chronic ozone exposure on net photosynthesis of various northeastern tree species. Exposures represent chronic-level ozone exposure (concentrations ≥ 150 ppb O_3). Numbers in parentheses correspond to numbers in reference listing.

Influence of Ozone on Carbon Allocation

Carbon Allocation

Elevated O_3 levels may alter patterns of carbohydrate allocation, either through changes in the various carbohydrate fractions or in the spatial allocation within the plant.

Rapid reductions in nonstructural carbohydrate contents of American elm (*Ulmus americana* L.) leaves, stems, and roots were recorded after short exposures to high O_3 concentrations.[18] Because the declines were proportional to the reduction in net photosynthesis, the changes were not indicative of altered allocation per se but were attributed to decreased photosynthesis. However, shifts in spatial allocation may have occurred also because carbohydrate content was reduced more in older leaves than in young expanding leaves. By the end of the fifth week following fumigation, carbohydrate levels of young leaves were comparable to those of nonfumigated control seedlings; the adverse effects on old leaves persisted, however. Long-term changes in content of reducing sugars, sucrose, and starch in roots of green ash (*Fraxinus pennsylvanica* Marsh.) seedlings exposed to O_3 were reported by Jensen.[19] Sucrose concentrations of fumigated seedlings showed more temporal variation than either the reducing sugar or starch contents (Figure 9). The authors hypothesized that the changes in carbohydrate concentration in the roots may be related to aboveground production. As the high demand for carbohydrates in the foliage and stems exceeded the supply available in the foliage, the seedlings mobilized starches and reducing sugars stored in the roots. Both of these carbohydrates would be transformed into sucrose before translocation, which would explain the variability in sucrose content. Dry matter accumulated in new stem and leaves more slowly in treated than in control plants, coinciding with the continued depletion of the carbohydrate reserves.

Using $^{14}CO_2$ to trace photosynthetically fixed carbon in foliage of white pine, Wilkinson and Barnes[20] found that exposure to 100 ppb O_3 resulted in a lower percentage of current photosynthate in the sugars and an increase in the sugar phosphates and amino acids (Figure 10). Conversely, growth in 200 ppb O_3 led to a decrease in the sugar phosphates, while the proportion of photosynthate in the organic acids increased. The authors concluded that the predominant O_3 effect was to divert photosynthetically fixed carbon from soluble sugars to amino and organic acids. Lower O_3 concentrations resulted in significant increases in total soluble carbohydrates, reducing sugars and ascorbic acid in primary needles of white pine.[21]

Belowground components of trees are dependent upon carbon allocation from photosynthesis. Root production and rhizosphere processes may be seriously impacted by decreases or shifts in carbon pools.

Kress and Skelly[22] examined effects of O_3 on growth of a number of eastern forest tree species. Root dry weight of sweetgum (*Liquidambar styraciflua* L.) and sycamore (*Platanus occidentalis* L.) was decreased significantly following O_3 exposures at 100 and 50 ppb O_3, respectively (Figure 11). Root biomass of sycamore was the most O_3-sensitive parameter of any of the species studied. Chappelka and Chevone[23] reported that root growth of white ash (*Fraxinus americana* L.) was adversely affected by O_3 at very low concentrations.

Root biomass of aspen growing under ambient O_3 conditions at a rural site in mid-state New York was not significantly different than that of seedlings growing in charcoal-filtered chambers.[24] However, aboveground shoot production was

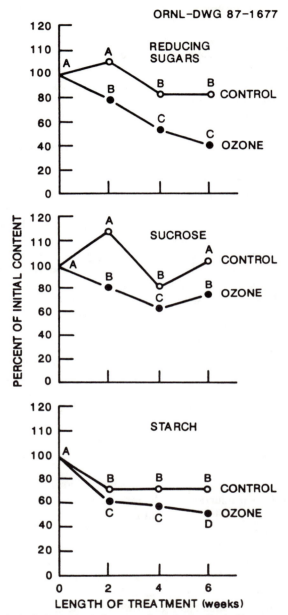

Figure 9. Carbohydrate content of green ash roots as affected by ozone fumigation (500 ppb O_3 for 8 h/day for 5 days/week). Treatments with the same letter are not significantly different at the 0.01 probability level. [Source: Jensen, K. F. "Ozone Fumigation Decreased the Root Carbohydrate Content and Dry Weight of Green Ash Seedlings," *Environ. Pollut.* 26:147–152 (1981).]

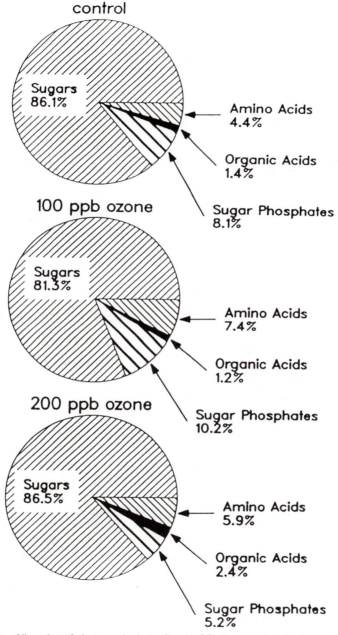

Figure 10. Allocation of photosynthetically fixed $^{14}CO_2$ in white pine seedlings as affected by ozone fumigation. [Source: Wilkinson, T. G., and R. L. Barnes. "Effects of Ozone on $^{14}CO_2$ Fixation Patterns in Pine," *Can. J. Bot.* 51:1573–1578 (1973).]

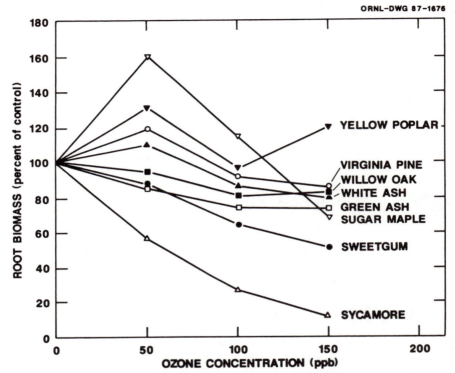

ORNL-DWG 87-1676

Figure 11. Root biomass as affected by ozone fumigation (6 h/day for 28 days). [Source: Kress, L. W., and J. M. Skelly. "Response of Several Eastern Forest Tree Species to Chronic Doses of Ozone and Nitrogen Dioxide," *Plant Disease* 66(12):1149–1152 (1982).]

lower for trees grown under ambient conditions. The root-to-shoot ratio did change with treatment, suggesting a relative increase in carbon allocation from shoots to roots.

Contrary to results of most studies, Taylor and others[25] reported that root dry weight of seedlings exposed to 120 ppb O_3 was greater than that of nonfumigated control seedlings. This increase in root biomass occurred without a statistically significant change in root-to-shoot ratio (i.e., no change in carbon allocation), though a trend of decreasing plant biomass was noted. Clearly, the effect of O_3 on dry matter partitioning and accumulation in the belowground tissues is complex, varying with season, nutrients, and water availability, and may be species-dependent. Generalizations across species are not appropriate.

Mycorrhizae are important fungus-root symbioses, without which many tree species could not survive under natural conditions.[26] Reich and coworkers[27] examined mycorrhizal infection (percentage of short roots infected by mycorrhizal fungi) of white pine and red oak exposed to O_3 concentrations ranging from 20 to 140 ppb and found no significant effect of O_3 applied three times per week. Mycorrhizae were responsive to O_3 exposures five times per week, suggesting that the mycorrhizae and/or the seedlings can recover from less frequent fumigations. At low dosages mycorrhizal infection increased; at higher concentrations the number of infected roots decreased. These changes were attributed to altered seedling carbon metabolism. For example, mycorrhizal fungi are dependent upon root exudates for establishment and carbohydrates for food.[26] If carbohydrate allocation to the roots is affected by O_3 fumigation, effects upon mycorrhizae may be expected.

Mycorrhizae may serve to protect roots from adverse effects of some forms of pollution.[28,29] Infection of paper birch (*Betula papyrifera* Marsh.) and white pine with the fungus *Pisolithus tinctorius* resulted in compensation for growth reductions caused by O_3.[30] This may have resulted from an increased sink strength of mycorrhizal roots, leading to increased translocation of assimilates to the roots.[31]

Reproduction

Reductions in reproductive processes may also occur as a consequence of O_3 fumigation. Ozone effects upon photosynthesis, carbon allocation, and water relations may affect reproductive ability. Coppicing by hybrid poplar (*Populus deltoides* X *trichocarpa*) decreased by as much as 50% in response to six weeks of fumigation with 150 ppb O_3.[32] Coppicing is known to be related to carbohydrate reserves. Also, flowering and reproduction require large amounts of carbohydrates.[33] If this carbon is not available, reproductive fitness may be diminished.

Shifts and declines in stored carbohydrates can have significant implications for tree growth. McLaughlin and coworkers[34] reported changing patterns of photoassimilate distribution among needles, shoots, and roots of white pine under chronic air pollutant stress. They hypothesized that reductions in needle longevity and size, increased respiratory demands, and altered allocation patterns led to a gradual decline in vigor of the trees. Moreover, spring growth of many tree species depends on carbohydrate reserves accumulated during the previous year.[35,36] If reserves are depleted or diverted to other sinks, growth rates may be reduced. Defense against pathogens and stresses also relies heavily on stored carbohydrates.[37,38] Depleted carbohydrate reserves could contribute to increased susceptibility to pathogen-related growth declines.

Relationship Between Ozone Air Quality and Tree Growth

Impacts of elevated O_3 levels on any critical physiological processes could lead to significant decreases in aboveground growth, productivity, and reproductive success of individual trees. In Figure 12, data are presented from a number of studies examining the growth response of northeastern forest tree species to O_3.

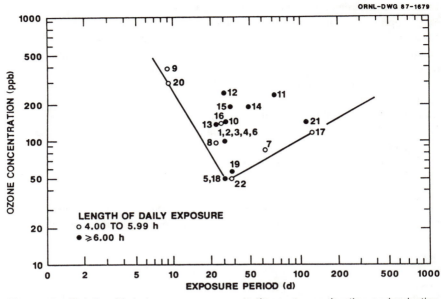

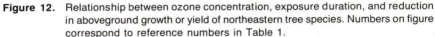

Figure 12. Relationship between ozone concentration, exposure duration, and reduction in aboveground growth or yield of northeastern tree species. Numbers on figure correspond to reference numbers in Table 1.

Coordinates represent experiments in which a statistically significant decrease, or a decline of $\geq 10\%$ in aboveground growth, was recorded due to O_3 exposure of known concentration and duration. Boundary line analysis was used to estimate exposure concentrations and durations resulting in significant decreases in growth and yield. Continuous O_3 concentrations above 50 ppb for exposure durations of >28 days result in decreased growth in the most sensitive northeastern species (e.g., poplar, sycamore, and sensitive genotypes of white pine). This threshold is twice the duration for sensitive agricultural species[9] and 50% longer than that for woody and herbaceous species from natural ecosystems representing all regions of North America. The distribution of the majority of coordinates in Figure 12 indicates that in order for most tree species to exhibit at least a 10% reduction in growth, the O_3 concentration and the duration of exposure must be substantially increased above the threshold for sensitive species. With longer exposure duration (>40 days), threshold concentrations appear to exceed 80 ppb. Based on the data in Figure 1, O_3 air quality in forested landscapes in the Northeast is, on the average, well below this threshold, suggesting that most tree species in the Northeast are not experiencing reductions in growth due to ambient levels of O_3.

Table 1.   Northeastern Tree Species Represented in Figure 12.

Reference Number	Tree Species	Reference Cited[1]
1	Liquidambar styraciflua	22
2	Platanus occidentalis	22
3	Fraxinus americana	22
4	Fraxinus pennsylvanica	22
5	Acer saccharum	22
6	Liriodendron tulipifera	22
7	Populus deltoides × trichocarpa	50
8	Fraxinus americana	23
9	Prunus serotina	48
10	Populus deltoides × trichocarpa	32
11	Betula papyrifera	51
12	Acer rubrum	52
13	Populus deltoides × trichocarpa	19
14	Acer saccharinum	53
15	Populus deltoides	54
16	Fraxinus americana	23
17	Picea rubens	25
18	Platanus occidentalis	55
19	Pinus strobus	30
20	Fraxinus americana	48
21	Populus deltoides × trichocarpa	56
22	Pinus strobus	49

[1]Numbers correspond with references at the end of the chapter.

INFLUENCE OF OZONE ON FOREST STANDS

The factors governing the response of forest stands to O_3 include the level of O_3 stress, seasonality of exposure, stand age, species composition, inter- and intraspecific competition, and the mix of site-specific abiotic and biotic stresses. Determination of O_3 as a causal factor in stand-level effects under field conditions is difficult because of the complexity inherent in working with large, dynamic natural systems and, usually, a mix of air pollutants. Many of the consequences of O_3 stress are manifestations of physiological responses of individual trees. For example, impaired photosynthesis and altered carbohydrate allocation could lead to slower growth or to less vigorous trees which, in turn, may be more susceptible to insects and other pathogens[39,40] or make them less competitive during periods of water or nutrient stress. Decreased pollen germination and seed production of individuals could lead to diminished reproductive success of sensitive species.

Decreases in stand biomass or productivity may occur if the growth of individual sensitive tree species is adversely affected. While the significance of elevated O_3

levels on the structure and function of selected arid forests in California has been characterized,[41] the extent to which O_3 affects forests in the northeastern United States is not well documented. Decreased growth and increased mortality of red spruce have been documented at several locations throughout the Appalachian Mountains.[42,43] McLaughlin and coworkers[44] reported that while climatic change may have contributed to the observed changes, the degree of radial growth suppression is greater than would be expected based on past growth-climate relationships. The authors hypothesized that the influence of either recent, unique combinations of climatic stresses or the interactive intervention of other regional-scale stresses, such as atmospheric pollution, may be contributing to decline in growth of this species. Evidence to link these declines with elevated O_3 levels is lacking, however. Growth declines could be of considerable significance in stands composed of only one (sensitive) species, as might be found in commercial plantings.

It is far too simplistic to extrapolate responses of single trees to stand-level responses, particularly for mixed-species, uneven-aged stands. Of particular importance is the modifying role of interspecific and intraspecific competition. While O_3 may directly affect an individual tree's growth or reproductive success, indirect effects may be further propagated as the individual is less able to compete for light, water, nutrients, etc. Decreased growth, vigor, competitiveness, and regeneration all could lead to increased mortality and subsequent changes in stand structure. However, decreased growth of one individual or species could be offset by improved growth of another due to lessened competition.

Sensitivity to O_3 damage varies with tree species. Table 2 ranks northeastern forest species by sensitivity to O_3 damage and rate of growth.[45] Fast-growing tree species seem to be more sensitive to O_3 damage than slow-growing species. Also, most of the species rated as sensitive are early successional species, and those rated resistant are typical of later successional stages. This may suggest that late successional communities are more resistant to O_3 stress. Elevated ozone levels could effectively accelerate succession, due to differential sensitivities. It has also been hypothesized that injury or disturbance of the dominant tree species could return the stand to an earlier successional stage.[42,46] Intermittent, chronic O_3 stress, such as is common in the northeastern United States, may lead to a shift in species composition of forest stands, but the nature and magnitude of these shifts are uncertain. However, many factors mitigate the response of tree species to elevated O_3 levels, so this generalization must be considered with some caution.

Environmental and stand conditions will alter the response of individual trees and forests to O_3 stress. For example, trees under mild moisture stress may be less susceptible to O_3 injury than trees with adequate moisture supply. High-elevation forests experience a variety of environmental stresses including acid cloud vapor, temperature extremes, and shallow soils. These stresses may either interact with O_3 directly or predispose these forests to O_3 effects.[39] Variations in climate and exposure add further to the uncertainty of O_3 effects.

Table 2.   Susceptibilities of Northeastern Tree Species to Ozone with Reference to Comparative Rates of Growth.

Resistant	Susceptible
Flowering dogwood — S (*Cornus florida* L.)	Green ash — R (*Fraxinus pennsylvanica* Marsh.)
Northern white cedar — S (*Thuja occidentalis* L.)	White ash — R (*Fraxinus americana* L.)
Sugar maple — S (*Acer saccharum* Marsh.)	Sweetgum — R (*Liquidambar styraciflua* L.)
Red oak — S (*Quercus rubra* L.)	Honey locust — R (*Gleditsia triacanthos* L.)
Black gum — R (*Nyssa sylvatica* Marsh.)	Pin oak — R (*Quercus palustris* Muenchh.)
Eastern hemlock — S [*Tsuga canadensis* (L.) Carr.]	Hybrid poplar — R (*Populus maximowiczii* × *trichocarpa*)
Black walnut — I (*Juglans nigra* L.)	Tulip poplar — R (*Liriodendron tulipifera* L.)
Red maple — I (*Acer rubrum* L.)	Sycamore — R (*Platanus occidentalis* L.)
American linden — S (*Tilia americana* L.)	Ailanthus — R [*Ailanthus altissima* (Mill.) Swingle]
Black locust — I (*Robinia pseudoacacia* L.)	Quaking aspen — R (*Populus tremuloides* Michx.)
	White oak — I (*Quercus alba* L.)

R = rapid growth rate, I = intermediate growth rate, S = slow growth rate.
Source: Harkov, R., and E. Brennan. "An Ecophysiological Analysis of the Response of Trees to Oxidant Pollution," *J. Air Pollut. Control Assoc.* 29:157–161 (1979).

A stochastic forest model (FORET) was used to simulate community dynamics in a pollutant-stressed forest of the southern Appalachians.[47] Growth rates of 32 tree species were reduced by 0, 10, and 20% based upon their assignment to resistant, intermediate, and sensitive categories, respectively. Changes in stand dynamics were projected for varying time periods, and species-specific tree growth rates were found to be modified by competitive interactions. While O$_3$ was not considered the primary pollutant, it is apparent that air pollution stress at the stand level may alter the distribution of species without affecting stand biomass.[48]

Changes in stand structure and function have implications for other forest ecosystem components. For example, nutrient quality of litter can vary with tree species. Thus, gradual shifts in species composition could affect soil nutrient pools and microfauna and flora, which depend upon these nutrient pools. Wildlife populations could be influenced by changes in mast production or the absence of suitable habitat if species are eliminated or diminished in importance.[9]

Under extreme pollutant stress, ecosystem collapse with associated soil erosion and irreversible loss of species from the system has been proposed.[46] However, Taylor and Norby[4] cautioned against such scenarios with respect to O_3, based on an understanding of the uniqueness of O_3 as an air pollutant in terrestrial ecosystems. In particular, O_3 is not selectively retained or accumulated in particular components of the ecosystem because of its highly reactive, ephemeral nature. Also, ecological effects are not restricted to areas immediately downwind of a major point source such as commonly occurs for emission of sulfur dioxides or heavy metals. Taylor and Norby concluded that the most likely effects of such a stress will be subtle, long-term shifts in species composition, rather than widespread community degradation.

CONCLUSIONS AND IMPLICATIONS

As an anthropogenic stress in natural ecosystems, O_3 exhibits many unique ecological and toxicological attributes including high reactivity, abbreviated residence time, regional distribution, and distinct spatial and temporal patterns in exposure dynamics. This uniqueness of O_3 as an environmental stress presents a significant problem in efforts to characterize the pollutant's effects on forest ecosystems because the conceptual framework and analytical methodologies for evaluating system-level responses are based on studies in acutely stressed landscapes. It is proposed that these methods are not sensitive enough to evaluate long-term ecosystem responses to a nonaccumulating, chronic level, regionally distributed stress such as O_3.

Methodologies for characterizing O_3 air quality are reviewed, with particular reference to forests in the northeastern United States. It is proposed that analyses be more ecologically and physiologically based. In particular, exposure statistics are recommended that (a) capture the temporal and spatial characteristics of O_3 air quality, (b) incorporate an understanding of the ecological properties of the pollutant and the physiological sensitivity of forest trees, and (c) recognize the pollutant's unique chemical attributes that make it distinct from other pollutants. With respect to forest ecosystems in the Northeast, the most distinctive feature of O_3 air quality is the dynamics of episodically high O_3 concentrations during the growing season.

Near-ambient O_3 concentrations are reported to affect some physiological processes in sensitive species and individuals. Carbon metabolism is particularly sensitive, and effects upon both carbon gain (photosynthesis) and carbon allocation have been documented. Declines or shifts in carbohydrate economy can significantly alter rhizosphere processes and dynamics, vigor, reproduction, and growth of trees. Individuals or species sensitive to O_3 may experience significant growth declines. With reference to northeastern forests, O_3 concentrations of 50 ppb or greater for a 28-day exposure period were found to affect growth and yield of sensitive forest species. However, the majority of species require substantially longer exposure or higher concentrations before growth reductions occur. In mixed stands, the most probable effects of O_3 stress will be subtle, long-term shifts in species composition, due to impacts on sensitive individuals or species.

ACKNOWLEDGMENTS

Authors acknowledge support from the Electric Power Research Institute under contract DE-AC05-840R21400 with Martin Marietta Energy Systems, Inc. and the U.S.D.A. Forest Service. The senior author was supported in part by an appointment to the Laboratory Cooperative Postgraduate Research Training Program administered by Oak Ridge Associated Universities. ESD Publication No. 3074, Environmental Sciences Division, Oak Ridge National Laboratory.

REFERENCES

1. Tingey, D. T., and G. E. Taylor, Jr. "Variation in Plant Response to Ozone, a Conceptual Model of Physiological Events," in *Effects of Gaseous Air Pollution in Agriculture and Horticulture,* M. H. Unsworth and D. D. Ormrod, Eds. (London: Butterworth, 1982), pp. 113–138.
2. Woodwell, G. M. "Effects of Pollution on the Structure and Physiology of Ecosystems," *Science* 168:429–433 (1970).
3. Lefohn, A. S., and C. K. Jones. "The Characterization of Ozone and Sulfur Dioxide Air Quality Data for Assessing Possible Vegetation Effects," *J. Air Pollut. Control Assoc.* 36(10): 1123–1129 (1986).
4. Taylor, G. E., Jr., and R. J. Norby. "The Significance of Elevated Levels of Ozone on Natural Ecosystems of North America," in *International Specialty Conference on Evaluation of the Scientific Basis for Ozone/Oxidant Standard,*" S. D. Lee, Ed. (Pittsburgh: Air Pollution Control Association, 1985), pp. 152–175.
5. Guderian, R., Ed. *Air Pollution by Photochemical Oxidants: Formation, Transport, Control, and Effects on Plants.* (Berlin: Springer-Verlag, 1985).
6. Taylor, G. E., D. T. Tingey, and H. C. Ratsch. "Ozone Flux in *Glycine max* (L.) Merr.: Sites of Regulation and Relationship to Leaf Injury," *Oecologia* (Berlin) 53:179–186 (1982).
7. Larcher, W., Ed. *Physiological Plant Ecology,* 2nd ed. (Berlin: Springer-Verlag, 1982).
8. Chambers, J. L., T. M. Hinckley, G. S. Cox, C. L. Metcalf, and R. G. Aslin. "Boundary-line Analysis and Models of Leaf Conductance for Four Oak-Hickory Forest Species," *For. Sci.* 31:437–450 (1985).
9. "Air Quality Criteria for Ozone and Other Photochemical Oxidants," Volume III. U.S. Environmental Protection Agency, EPA/600/8-84/020cF (1986).
10. Levitt, J. *Responses of Plants to Environmental Stresses* (New York: Academic Press, 1972).
11. Prinz, B., G. H. M. Krause, and K.-D. Jung. "Neuere Untersuchungen der LIS zu den neuartigen Waldschaden," *Dusseldorfer Geobot. Kollog.* 1:25–36 (1984).

12. Lindberg, S. E., D. Silsbee, D. A. Schafer, J. G. Owens, and W. A. Petty. "Comparison of Atmospheric Exposure Conditions at High-elevation and Low-elevation Forests in the Southern Appalachian Mountains," in *Acid Deposition at High Elevations*. M. H. Unsworth and D. Fowler, Eds. (Norwell, MA: Kluwer Academic Publishers, 1988), pp. 321–344.

13. Reich, P. B., and R. G. Amundson. "Ambient Levels of Ozone Reduce Net Photosynthesis in Tree and Crop Species," *Science* 230:566–570 (1985).

14. Boyer, J. N., D. B. Houston, and K. F. Jensen. "Impacts of Chronic SO_2, O_3, and $SO_2 + O_3$ Exposures on Photosynthesis of *Pinus strobus* Clones," *Eur. J. For. Path.* 16(5–6):293–299 (1986).

15. Tingey, D. T., and W. E. Hogsett. "Water Stress Reduces Ozone Injury via a Stomatal Mechanism," *Plant Physiol.* 77(4):944–947 (1985).

16. McLaughlin, S. B., and G. E. Taylor, Jr. "Relative Humidity: Important Modifier of Pollutant Uptake by Plants," *Science* 221:167–169 (1981).

17. Norby, R. J., and T. T. Kozlowski. "The Role of Stomata in Sensitivity of *Betula papyrifera* Seedlings to SO_2 at Different Humidities," *Oecologia* 53:34–39 (1982).

18. Constantinidou, H. A., and T. T. Kozlowski. "Effects of Sulfur Dioxide and Ozone on *Ulmus americana* seedlings. II. Carbohydrates, Proteins, Lipids," *Can. J. Bot.* 57:176–184 (1979).

19. Jensen, K. F. "Ozone Fumigation Decreased the Root Carbohydrate Content and Dry Weight of Green Ash Seedlings," *Environ. Pollut.* 26:147–152 (1981).

20. Wilkinson, T. G., and R. L. Barnes. "Effects of Ozone on $^{14}CO_2$ Fixation Patterns in Pine," *Can. J. Bot.* 51:1573–1578 (1973).

21. Barnes, R. L. "Effects of Chronic Exposure to Ozone on Soluble Sugar and Ascorbic Acid Contents of Pine Seedlings," *Can. J. Bot.* 50:215–219 (1972).

22. Kress, L. W., and J. M. Skelly. "Response of Several Eastern Forest Tree Species to Chronic Doses of Ozone and Nitrogen Dioxide," *Plant Disease* 66(12):1149–1152 (1982).

23. Chappelka III, A. H., and B. I. Chevone. "White Ash Seedling Growth Response to Ozone and Simulated Acid Rain," *Can. J. For. Res.* 16(4):786–790 (1986).

24. Wang, D., D. F. Karnosky, and F. H. Bormann. "Effects of Ambient Ozone on the Productivity of *P. tremuloides* Michx. Grown Under Field Conditions," *Can. J. For. Res.* 16:47–55 (1986).

25. Taylor, G. E., Jr., R. J. Norby, S. B. McLaughlin, A. H. Johnson, and R. S. Turner. "Carbon Dioxide Assimilation and Growth of Red Spruce (*Picea rubens* Sarg.) Seedlings in Response to Ozone, Precipitation Chemistry, and Soil Type," *Oecologia* 70:163–171 (1986).

26. Kormanik, P. P., W. C. Bryan, and R. C. Schultz. "The Role of Mycorrhizae in Plant Growth and Development, in *Physiology of Root-Microorganisms Associations*, H. Max Vines, Ed. South. Section. Am. Soc. Plant Physiol., Atlanta (1977).

27. Reich, P. B., A. W. Schoettle, and R. G. Amundson. "Testing for Interactions Between Acid Rain and Ozone: Response of Photosynthesis, Mycorrhizae, Plant Nutrition, and Growth in White Pine." Ninth North American Forest Biology Workshop, Stillwater, OK, June 15–18, 1986.

28. Carney, J. L., H. E. Garnett, and H. G. Hedrick. "Influence of Air Pollution Gases on Oxygen Uptake of Pine Roots with Selected Ectomycorrhizae," *Phytopathology* 73:1035–1040 (1978).

29. Garrett, H. E., J. L. Carney, and H. G. Hedrick. "Influence of Ozone and Sulfur Dioxide on Respiration of Ectomycorrhizal Fungi," *Can. J. For. Res.* 12:141–145 (1982).

30. Keane, K. D., and W. J. Manning. "Ozone and Acidic Precipitation: Effects on the Growth of Mycorrhizal and Non-Mycorrhizal Paper Birch and White Pine Seedlings," *Phytopathology* 76(6):654 (1986).

31. Handley, W. R. C., and C. I. Sanders. "The Concentration of Easily Soluble Reducing Substances in Roots and the Formation of Ectotrophic Mycorrhizal Associations: A Reexamination of Bjorkman's Hypothesis," *Plant Soil* 16:42–61 (1962).

32. Jensen, K. F., and L. S. Dochinger. "Response of Hybrid Poplar Cuttings to Chronic and Acute Levels of Ozone," *Environ. Pollut.* 6:289–295 (1974).

33. Kramer, P. J., and T. T. Kozlowski. *Physiology of Woody Plants* (New York: Academic Press, Inc., 1979).

34. McLaughlin, S. B., R. K. McConathy, D. N. Duvick, and L. K. Mann. "Effects of Chronic Air Pollution Stress on Photosynthesis, Carbon Allocation, and Growth of White Pine Trees," *For. Sci.* 28:60–70 (1982).

35. Ericsson, A., and H. Persson. "Seasonal Changes in Starch Reserves and Growth of Fine Roots of 20-Year-Old Scots Pine," in *Structure and Function of Northern Coniferous Forests—An Ecosystem Study,* T. Persson, Ed. *Ecol. Bull.* (*Stockholm*) 32:239–250 (1980).

36. Ford, E. D., and J. D. Deans. "Growth of a Sitka Spruce Plantation: Spatial Distribution and Seasonal Fluctuations of Lengths, Weights, and Carbohydrate Concentrations of Fine Roots," *Plant Soil* 47:463–485 (1977).

37. Webb, W. L. "Relation of Starch Content to Conifer Mortality and Growth Loss after Defoliation by the Douglas-fir Tussock Moth," *For. Sci.* 27(2):224–232 (1981).

38. McLaughlin, S. B., and D. S. Shriner. "Allocation of Resources to Defense and Repair," in *Plant Disease,* J. G. Horsfall and E. B. Cowling, Eds. (New York: Academic Press, 1980), p. 53.

39. McLaughlin, S. B. "Effects of Air Pollution on Forests: A Critical Review," *J. Air Pollut. Control Assoc.* 35:512–534 (1985).

40. Woodman, J. N., and E. B. Cowling. "Airborne Chemicals and Forest Health," *Environ. Sci. Technol.* 21(2):120–126 (1987).

41. Miller, P. R., O. C. Taylor, and R. G. Wilhour. "Oxidant Air Pollution Effects on a Western Coniferous Forest Ecosystem." U.S. Environmental Protection Agency EPA-600/D-82-276 (1982).

42. Siccama, T., M. Bliss, and H. W. Vogelman. "Decline of Red Spruce in the Green Mountains of Vermont," *Bull. Torrey Bot. Club* 109:162–168 (1982).

43. Johnson, A. H., and T. G. Siccama. "Acid Deposition and Forest Decline," *Environ. Sci. Technol.* 17:294A–305A (1983).

44. McLaughlin, S. B., D. J. Downing, T. J. Blasing, E. R. Cook, and H. S. Adams. "An Analysis of Climate and Competition as Contributors to Decline of Red Spruce in High Elevation Appalachian Forests of the Eastern United States," *Oecologia* 72:487–501 (1987).

45. Harkov, R., and E. Brennan. "An Ecophysiological Analysis of the Response of Trees to Oxidant Pollution," *J. Air Pollut. Control Assoc.* 29(2):157–161 (1979).

46. Bormann, F. H. "Air Pollution and Forests: An Ecosystem Perspective," *BioScience* 35:434–441 (1985).

47. West, D. C., S. B. McLaughlin, and H. H. Shugart. "Simulated Forest Response to Chronic Air Pollution Stress," *J. Environ. Qual.* 9:43–49 (1980).
48. McClenahen, J. R. "Community Changes in a Deciduous Forest Exposed to Air Pollution," *Can. J. For. Res.* 8:432–438 (1978).
49. Yang, Y. S., J. M. Skelly, B. T. Chevone, and J. B. Birch. "Effects of Long-Term Ozone Exposure on Photosynthesis and Dark Respiration of Eastern White Pine," *Environ. Sci. Technol.* 17(6):371–373 (1983).
50. Reich, P. B., and J. P. Lassoie. "Influence of Low Concentrations of Ozone on Growth, Biomass Partitioning, and Leaf Senescence in Young Hybrid Poplar Plants," *Environ. Pollut.* (Series A) 39:39–51 (1985).
51. Jensen, K. F., and R. G. Masters. "Growth of Six Woody Species Fumigated With Ozone," *Plant Dis. Rep.* 59:760–762 (1975).
52. Dochinger, L. S., and A. M. Townsend. "Effects of Roadside Deicer Salts and Ozone on Red Maple Progenies," *Environ. Pollut.* 19:229–237 (1979).
53. Jensen, K. F. "Growth Relationships in Silver Maple Seedlings Fumigated with O_3 and SO_2," *Can. J. For. Res.* 12:298–302 (1983).
54. Jensen, K. F. "An Analysis of the Growth of Silver Maple and Eastern Cottonwood Seedlings Exposed to Ozone," *Can. J. For. Res.* 12:420–424 (1982).
55. Kress, L. W., J. M. Skelly, and K. H. Hinkelmann. "Growth Impact of O_3, NO_2, and SO_2 on *Platanus occidentalis*," *Agric. Environ.* 7:265–274 (1982).
56. Patton, R. L. "Effects of Ozone and Sulfur Dioxide on Height and Stem Specific Gravity of *Populus* Hybrids." USFS Research Paper NE-471. Northeastern Forest Experiment Station (1981).

CHAPTER 6

Effects of Ozone on Tires and the Control of These Effects

Joseph A. Kuczkowski

INTRODUCTION

Atmospheric ozone degrades polydiene rubbers by reacting directly with polymer main chain unsaturation. This causes chain scission and reduced surface strength which is grossly manifested by the appearance of cracks on the surface of the vulcanized rubber product.[1] Rubbers thus affected include natural rubber (NR), synthetic *cis*-polyisoprene (IR), styrene-butadiene rubber (SBR), polybutadiene rubber (BR), and acrylonitrile-butadiene rubber (NBR). Other rubbers such as ethylene-propylene-diene terpolymer (EPDM) or chlorobutyl rubber (XIIR) which have highly saturated backbones react very slowly with ozone and do not show this cracking.[2] Ozone cracking characteristically occurs in a pattern which is perpendicular to some applied external or internal stress. The cracking is a direct result of chain scission which produces several different oxygen-containing decomposition products.[1] The mechanism follows the format below, which was originally suggested by Criegee.[3]

Some of the observed reaction products include the primary ozonide (molozonide) (1), carbonyl oxide (zwitterion) (2), aldehydes and ketones (3), secondary ozonide (4), polymeric peroxides (5), and hydroperoxide species (6).

Research on ozone induced crack growth has been an active area of study for over 25 years.[4-13] These studies established (a) that the rate of ozone absorption for a typical diene polymer is linear and directly proportional to the ozone concentration, and (b) that unstressed rubber still reacts with ozone to form oxidized

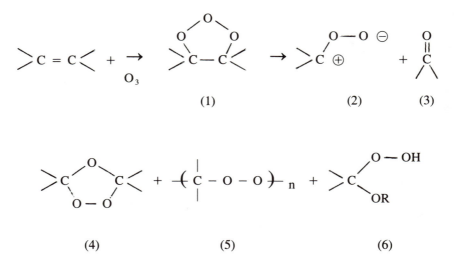

(1)   (2)   (3)

(4)   (5)   (6)

surface films even though it does not show the characteristic cracking patterns.[14] The work of Braden and Gent identified two characteristic parameters which are necessary for the formation and propagation of cracks under static strain conditions.[4,5] The first is that growth will not occur unless a minimum tearing energy is exceeded by stretching the rubber. This has become known as the "critical stress" effect. For diene rubbers, it is generally of similar value and corresponds to a threshold tensile strain of 3–5%. The second key observation was that once the "critical stress" was excceded, a single isolated crack growth was independent of the strain and directly proportional to the ozone concentration under the conditions of (a) described above.

The type of cracks that will form in a polymer surface, once the "critical stress" is exceeded, is variable. For rubbers which are just slightly above this value, the cracks are generally fewer but larger and more severe. For rubbers which are under much higher strain, the cracks are more numerous and less severe. At very high strain, sometimes these cracks are so fine as to become microscopic.[15] This difference in crack type is due to differences in rate of crack initiation, and is also a reflection that cracking is a surface stress relaxation process.

CHEMICAL ANTIOZONANTS

Today rubber vulcanizates are protected through the use of chemical antiozonants. These protectants possess several common features: (a) all effective materials react competitively and directly with ozone; (b) they must migrate to the surface of the rubber to be effective; (c) their presence decreases the rate of cut growth; (d) some materials such as the N,N'-dialkyl-p-phenylenediamines and the N-alkyl-N'-aryl-p-phenylenediamines also affect the polymer by raising the apparent critical stress.[13]

One additional important discovery relates to the actual production of atmospheric ozone. This process is directly dependent upon the photoenergy contained in sunlight. Surface ozone concentration increases in the daytime and decreases through the nighttime in a daily repetitive pattern. The production of atmospheric ozone by sunlight depends on several different factors. Among these are the geographic location, the climate, the season, the weather, and air pollution. Since migration of a chemical antiozonant to the surface of a rubber product occurs at a rather constant rate, the nighttime relaxation time allows a replenishment of surface protectant for the following day's peak exposure.[16]

The need for protecting rubber products from the degrading effects of ozone was not actually realized until after the introduction of antioxidants such as N-phenyl-*beta*-naphthylamine and N,N'-diphenyl-p-phenylenediamine in the 1930s.[17,18] The use of these stabilizers, along with the introduction of new organic vulcanization accelerators, led to a threefold increase of tire lifetime.[19] These antioxidant materials provided for oxidative stability, but they did nothing to provide for dynamic ozone protection. Typical service lifetime at that time was extended from 3,500 miles or one year of service to over 10,000 miles or two years. Continued improvements in tire quality and construction have produced a product which today will give the consumer 50,000 miles or three to four years of relatively trouble-free service.[20]

Early attempts to provide ozone protection centered on materials which would form barrier films by exuding or blooming to the polymer surface. Various tars and pitches were tried. The best degree of static ozone protection was offered through the use of waxes of both the paraffin and microcrystalline type. These types of waxes successfully bloom to the surface of a rubber article and form an inert surface barrier film. Since waxes are essentially pure alkane hydrocarbons which contain no unsaturation, they do not react with ozone and, therefore, they provide only a physical barrier to ozone attack. This physical protection fails under flexing conditions as the barrier film is broken. Therefore, chemically reactive antiozonants were sought and developed to meet this dynamic protection need.[21,22]

One of the first chemical antiozonant systems discovered for rubber was based on a derivative of dihydroquinoline. This was 1,2-dihydro-6-ethoxy-2,2,4-trimethylquinoline (DETQ).[23]

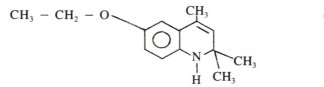

(7)

DETQ provided adequate protection at the time of its use; however, it had several annoying properties. Among these were that it was quite volatile and it caused severe staining and discoloration of white wall tires as well as adjacent automotive painted areas.[1] Polymerized derivatives of quinoline were also tried but did not migrate as well and, therefore, provided only mild antiozonant activity.[24,25]

Continued investigations into the use of amine based materials as antiozonants led to the use of the gasoline antioxidant N,N'-di-*sec*-butyl-p-phenylenediamine.

$$\text{(8)}$$

This material was very effective as an antiozonant for vulcanized rubber, but it exhibited all of the same detrimental properties of DETQ and, in addition, was found to be very toxic. Failure of rubber parts on military vehicles which were stored from World War II and then pressed into service for the Korean War encouraged the government to support a systematic study of the p-phenylenediamines as antiozonants for rubber. This became known as the Rock Island Arsenal study.[26] From this study, it was determined that a broad class of N,N'-(di-*sec*-alkyl)-p-phenylenediamines was most effective as antiozonants.[27] An additional generalization showed that the oxidation potentials for antiozonants fall in a range which is grossly different and lower than that noted for several established amine antioxidant systems. A proper oxidation potential, however, is not sufficient to guarantee practical utility. Materials such as the p-aminophenols and N,N'-di-(primary-alkyl)-p-phenylenediamines did have the appropriate oxidation potentials but still provide little or no ozone protection. Failure of these systems is due to competitive destructive reactions with oxygen and also with the vulcanization chemistry involved in the curing process.[28]

Several dioctyl-p-phenylenediamines became commercially available after the Rock Island study. These materials are much less toxic than the dibutyl analog, and the presence of the longer chain alkyl groups markedly reduces the volatility. The major materials marketed were N,N-*bis*-(1-ethyl-3-methylpentyl)-p-phenylenediamine (88PD) and N,N'-(1-methylheptyl)-p-phenylenediamine (DOPPD). These materials were prepared respectively by reductive alkylation of p-phenylenediamine with 2-octanone or 5-methyl-3-heptanone.[29] Although these materials are highly active, their practical use in rubber is short-lived. This is due to the fact that these materials also react readily with oxygen to produce other transformation products. Such reactions are detrimental and reduce the concentration of the effective antiozonant.[28]

After the introduction of the dialkyl-p-phenylenediamines, research continued on the development of other related systems. Alkyl/aryl- analogs such as N-isopropyl-N'-phenyl-p-phenylenediamine (IPPD) and N-cyclohexyl-N'-phenyl-p-phenylenediamine were then made commercially available.[30] This was followed by the introduction of a mixed diaryl- system, N,N'-diaryl p-phenylenediamine

(DAPD).[31] Both the N-cyclohexyl-N'-phenyl- and the N,N'-diaryl- derivatives provided a partial solution to the lack of persistence noted for the dialkyl-derivatives. However, the antiozonant activity of the two new systems was somewhat less. The IPPD system maintains the high activity of the dialkyl- system, but it is quite volatile and is easily extracted from the rubber vulcanizate through contact with acid rain.[32] A further compromise was found in N-(1,3-dimethyl-butyl)-N'-phenyl-p-phenylenediamine (6PPD) and higher alkyl-derivatives based on C_7 and C_8 alkyl groups.[30] The 6PPD is highly active and not easily extracted. Today, the use of 6PPD in the United States is far greater than that of all the other phenylenediamines combined.

INTERACTIVE PROPERTIES OF ANTIOZONANTS

The p-phenylenediamines as a class have additional beneficial properties for use in rubber other than their activity as antiozonants. These materials also provide polymer antioxidant protection, serve as metal deactivators, provide beneficial stress relaxation and improve rubber wear and resilience. The diaryl-p-phenylenediamines have replaced N-phenyl-*beta*-naphthylamine as strong antioxidants for vulcanizates and also as raw polymer stabilizers.[31] All materials of this type are deficient in that they cause compound stain and discoloration. Their prime use is, therefore, limited to carbon black loaded vulcanizates.[33]

All p-phenylenediamine antiozonants have an accelerating effect on the rubber vulcanization system. This is due to the fact that they are nitrogenous bases. The intensity of this effect is a direct function of the amines basicity and it can decrease the scorch or process safety time for a compound. The dialkyl- materials are more accelerating or scorchy than the alkyl/aryl- derivatives, while the diaryl-p-phenylene-diamines show the least effect.[29,33]

Various rubbers require differing degrees of ozone protection. Rubbers with saturated backbones (EPDM) require no protection at all, while rubbers such as XIIR with low amounts of unsaturation require minimal protection. Neoprene (CR) is much less susceptible to ozone attack than is natural rubber (NR). This is explained on the basis of the electron withdrawing effect that the chlorine atom has on the neoprene backbone unsaturation. Rubbers such as polybutadiene (BR) or styrene-butadiene are also easier to protect than is polyisoprene (IR), which has an electron rich double bond. Nitrile rubber (NBR) is the most difficult to protect against ozone and no chemical antiozonant system is totally satisfactory.[34]

The trivial solution to the ozone problem would be to use an inherently resistant rubber. Thus, replacement of the polydiene rubber with a saturated backbone rubber such as ethylene-propylene-diene terpolymer (EPDM) should avoid the problem. Such an approach is naive, however, since various rubbers possess other physical properties which cannot be maintained in the substitution. Among some of these are resilience, rebound, tear strength, heat buildup, abrasion, and ease of processing and vulcanization.[2]

The degree of required ozone protection is also a function of several additional factors. Good component design cannot be neglected. Smooth vulcanizates need less protection than those molded into sharply defined geometric shapes. Rubber vulcanizates which are unstressed will not show ozone cracks. The effectiveness of an antiozonant can also be increased through the combined use of antioxidants and waxes. This can reduce the amount of antiozonant needed to provide adequate protection. The antioxidants are thought to protect the antiozonant system from direct oxidation while the waxes protect the rubber during static rest periods. Unusual surface stresses can also be obtained from materials which bloom from the rubber compound as solid crystals. Such materials should be avoided since they serve as starter points for ozone attack.[1,2,34]

For long term protection, the compounder must also be aware of what changes in antiozonant activity will occur with length of service. While dialkyl-p-phenylenediamines are initially the most active antiozonants, followed by the alkyl/aryl-, and then the diaryl- derivatives, this order of activity reverses itself as the system is aged due to oxidation and exposure to ozone.[35,36] For extended protection, mixtures of these materials are often used.[37,38]

The polymer matrix must also be considered in the choice of the antiozonant. For pure NR vulcanizates, the effective use of diaryl-p-phenylenediamine is diminished due to solubility problems. Less than one part per hundred is soluble and this is below the amount needed for protection. However, NR is used frequently in combination with other rubbers. These blends allow effective amounts of the diaryl- system to be used. This solubility phenomenon is less of a problem in rubbers such as BR or SBR. The difficulties encountered in protecting NBR are partially due to the high solubility of the p-phenylenediamines in this polymer. High solubility means insufficient amounts migrate to the surface. In neoprene (CR), the order of antiozonant activity actually reverses itself. The diaryl-p-phenylenediamines are much more persistent and effective than the dialkyl- or alkyl/aryl- systems.

In addition to the p-phenylenediamines and the substituted trimethyldihydroquinolines, other chemical antiozonant systems have been investigated and commercialized. Among these are systems based on N-substituted thioureas, phosphites, substituted olefins and nickel dialkyldithiocarbamates. In general, all of these systems have much less activity and utility than the p-phenylenediamines, the one exception being nickel dibutyldithiocarbamate. Use of this system can cause problems since it is based on a toxic heavy metal. It also turns the vulcanizate green in color and it is prooxidative in natural rubber.[39]

While the p-phenylenediamines have primary problems with color and staining, these other systems also have their unique problems and advantages. The N-substituted thioureas are moderate to good nonstaining nondiscoloring antiozonants. However, their practical use is limited since they cause premature vulcanization and cure interference problems.[40-43] Although phosphites are active in unvulcanized stock, their activity is lost on sulfur or peroxide vulcanization as these materials are converted to inactive thiophosphate or phosphate esters.[44] Substituted olefins are typically used in rubbers which are already some-what ozone resistant, e.g., neoprene (CR). These materials are nonstaining and function in a sacrificial fashion by offering a more ozone-reactive double bond than is contained in the polymer.[45]

THEORIES OF ANTIOZONANT PROTECTION

As a stressed rubber surface is ozonized, minute cracks begin to form. The primary ozonide (1) cleaves under this stress to form carbonyl oxide (2) and ketonic fragments (3). Intermediates (2) and (3) need not recombine with themselves to form the secondary ozonide (4), but may do so with nearest similar neighboring groups. The macroscopic effect is that stress is relaxed at differential segments of the surface and a crack begins to form. Stress which was originally at the sur-face is then transferred to the point of crack growth. This creates a new region of high stress and ozone attack continues in this area to propagate the cleavage. A model for this process is below.

$\longleftarrow$ **RELEASE OF SURFACE STRESS** $\longrightarrow$

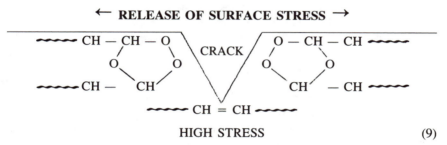

HIGH STRESS   (9)

Although several chemical antiozonant systems are in practical use, the exact mechanism of their protection is still not fully understood. This area continues to be one of extensive active research. Various theories have been proposed to account for this mechanism of antiozonant protection.[46] These can be categorized into the following four functional theoretic models.

a. inert barrier
b. competitive reaction
c. reduced critical stress
d. chain repair

The inert barrier theory proposes that the antiozonant migrates from the bulk of the rubber to the surface to form a film. This film is a physical barrier which protects reactive polymer double bonds by keeping ozone out of contact. This is the mechanism by which waxes are believed to function. It can also be used to describe the ozone resistance obtained for blends of diene polymers with non-reactive polymers such as EPDM or XIIR. It also explains the ozone resistance obtained by post reacting diene rubber surface with halogen, oxygen, or mercaptans.[34]

The competitive reaction theory can be subdivided into two parts. First, the ''scavenger'' model proposes that as the antiozonant migrates to the surface, it undergoes selective reactions with ozone. This reaction selectivity protects the polymer double bonds until the antiozonant is exhausted.[28] Then the ''protective film'' theory proposes that once the antiozonant has functioned as a ''scavenger,'' the transformation products of the reaction form an inert protective film.[14] This dual function appears to be a common feature for all chemical antiozonants.

A considerable amount of experimental data supports the ''scavenger'' mechanism as being dominant. This conclusion comes from data collected from various sources and under different research approaches. These findings will be summarized here. Ozone has been shown to react preferentially with a chemical antiozonant while in the presence of a solution of rubber. A screening test for antiozonant activity known as dilute solution viscosity is based on this observation.[47] Measurements of DOPPD diffusion to the rubber surface have shown that the rate is sufficient to ensure protection against nominal amounts of ozone.[48] Attenuated total reflectance infrared has shown that ozone preferentially attacks the antiozonant rather than the rubber surface.[49,50] Ozonation transformation products of DOPPD and 6PPD have been characterized individually and on the rubber surface. These products include amides, nitroso, and nitro compounds, nitrones and benzoquinonediimines.[51,52]

The formation of a ''protective film'' on the rubber surface is also rather well substantiated. Such films have been detected microscopically and visually with the naked eye. If the film is partially washed from the rubber surface and the sample reexposed to ozone, only the cleaned surface undergoes ozone degradation.[48] Analysis of the film by spectroscopic examination shows the presence of unreacted antiozonant as well as several transformation products. These transformation products are essentially the same as those obtained by the direct ozonation of the antiozonant in solution. These polar transformation products apparently do not diffuse back into the rubber but remain on the surface. Attempts to add ozonized transformation products of an antiozonant back into a freshly compounded rubber have shown that these materials offer very little ozone protection to the new vulcanizate. The ozonized film, therefore, functions as a secondary chemical scavenger and physical barrier to ozone attack. The formation of this transformation product film is believed to be of secondary importance to the ''scavenger'' function.[34]

The reduced critical stress theory proposes that the rubber surface is modified by the migration of the antiozonant to or just below the surface. This modification relieves the internal and surface stresses so that the surface behaves as if it were unstressed rubber. As a result, no cracks form.[46,53] As previously mentioned, this property appears to be unique for the dialkyl- and alkyl/aryl-p-phenylenediamines. This modification of critical stress phenomenon is the least understood of all the theories, but the easiest to demonstrate. If rubber test samples containing various levels of an antiozonant are stressed from 0 to 100% and then exposed to ozone, a mapping of the crack formation and type will result. Typically, this test shows that higher levels of antiozonant will protect a rubber to a higher level of critical stress before cracking is noted. This phenomenon has formed the basis of an annulus test for critical stress modification.[54]

The chain repair theory proposes that antiozonants serve as "relinking" agents for severed polymer chains. This relinking occurs through reaction with end groups such as aldehydes or acids.[55,56] This theory also has been modified to suggest that the antiozonant reacts directly with the ozonide (4) or the carbonyl oxide (2), thus giving a low molecular weight, inert "self healing film."[57] Either of these proposals would result in the attachment of the antiozonant to the rubber. The bulk of the evidence in support of this theory has been gathered by doing extraction studies on the rubber, followed by nitrogen analysis of the rubber. Considerable amounts of nitrogen have been found to remain on the polymer after extraction. However, it is impossible to attribute this residual nitrogen as being entirely due to the above reactions. Other explanations for the bound nitrogen are also possible. Among these are transformations of the antiozonant that are due to interaction with oxygen and also with curative intermediates. The p-phenylene-diamines are known to be free radical traps and accelerators of vulcanization.[58,59] A recent additional suggestion proposes that exposure to oxygen or ozone can form a nitroxyl form of the antiozonant. This should result in direct attachment to the rubber.[60] Contributions from the "chain repair" or "self healing film" are regarded as tenuous. Of all the proposals, these may be the least important in the actual mechanism of antiozonant protection.[34,61]

The results of a study on the direct ozonation of DOPPD and 6PPD has been reported by the B.F. Goodrich Group.[50,52] A schematic of the types of transformation products which are produced from 6PPD is shown in Figure 1. Additional discussions on the role and interactions of transformation products of p-phenylenediamines can be found in a review by Pospíšil.[34]

CONCLUSIONS

In conclusion, a brief summary of those general properties that are necessary or desirable for a good rubber antiozonant is included. A good antiozonant must be very reactive with ozone. It must do so selectively and competitively with the carbon-carbon double bonds that are present in the rubber polymer. Materials which fail in this regard will not provide adequate protection against ozone. Conversely, it must not be too reactive with ozone, or it will not be persistent enough to provide a sufficient length of service lifetime. Likewise, the material should not be prooxidative through rapid direct reactions with oxygen. Both of these destructive interactions will severely limit its utility. The system should have no significant effects on the vulcanization and curative package. Detrimental reactions with sulfur or other rubber accelerators can deactivate a system. The solubility

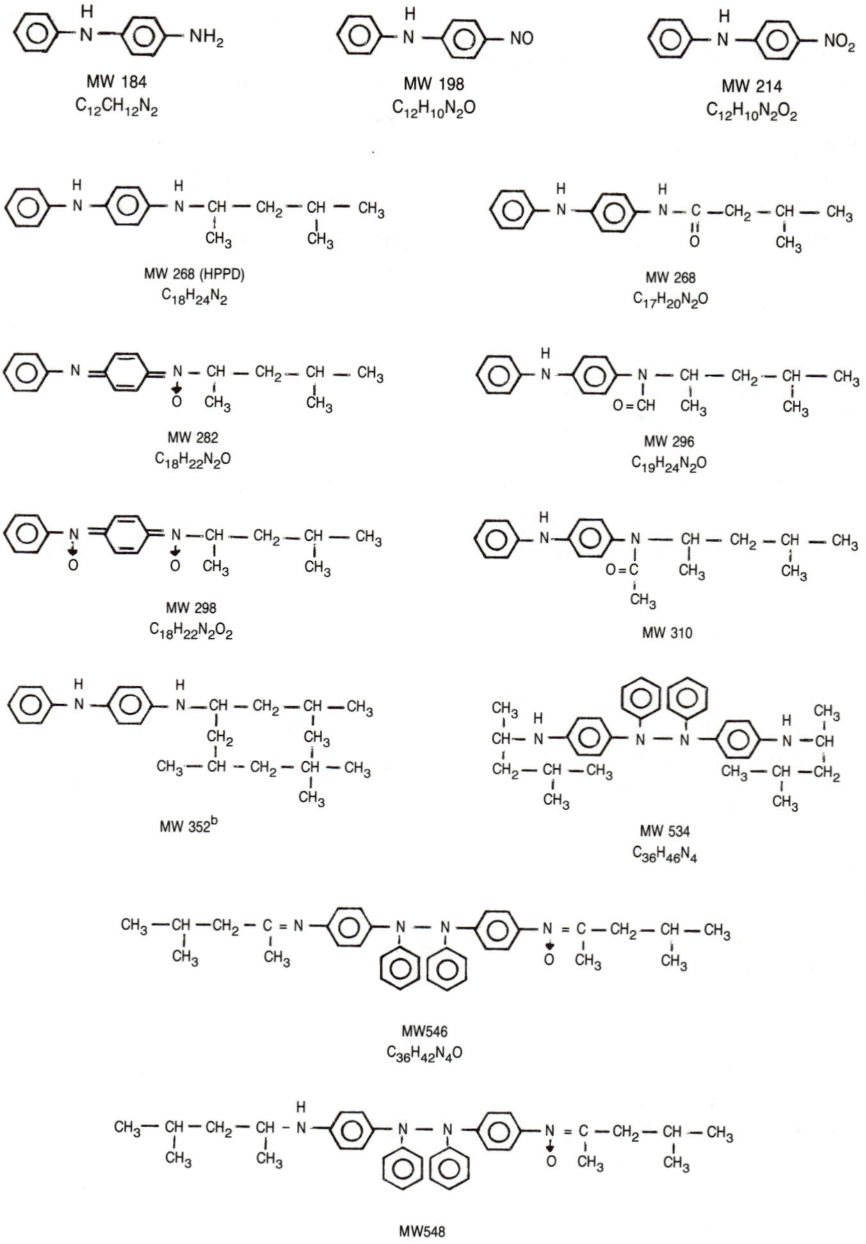

[a]Based on mass spectrometric data. Elemental formula (when given) was determined by atomic composition mass spectrometry.

[b]Impurity in starting material.

Figure 1. Structure for Ozone-6PPD reaction products[a]. Reproduced from Reference 52 with permission of the editor.

and diffusivity characteristics should allow it to migrate to the rubber surface at a linear rate which should match the amount needed for protection. Materials which have poor solubility often spew from the rubber and form annoying cosmetic blooms. On the other hand, excellent antiozonants which are too soluble will not migrate well and insufficient amounts will come to the surface to provide adequate protection. Once at the surface, the persistence and volatility properties should be such that the material remains on the surface. The ideal antiozonant should not discolor or cause staining problems. This ideal has yet to be attained satisfactorily. As a result, research is still actively being conducted in an effort to find a highly active, nonstaining, nondiscoloring antiozonant.

REFERENCES

1. Cox, W. L. "Antiozonants" in *Encyclopedia of Polymer Science and Technology* Vol. 2, H. F. Mark, Ed. (New York: John Wiley & Sons, Inc.. 1965), p. 197.
2. Lewis, P. M., *Polym. Degradation Stab.* 15:33 (1986).
3. Criegee, R., A. Kerckow, and H. Zinke, *Ber.* 88:1878 (1955).
4. Braden, M., and A. N. Gent. *J. Appl. Polym. Sci.* 3:90 (1960).
5. Braden, M., and A. N. Gent. *J. Appl. Polym. Sci.* 3:100 (1960).
6. Braden, M., and A. N. Gent. *Trans. Inst. Rubber Ind.* 37:88 (1961).
7. Andrews, E. H., and M. Braden. *J. Polym. Sci.* 55:787 (1961).
8. Braden, M., and A. N. Gent. *Rubber Chem. Technol.* 35:200 (1982).
9. Braden, M., and A. N. Gent. *J. Appl. Polym. Sci.* 6:449 (1962).
10. Andrews, E. H., and M. Braden. *J. Appl. Polym. Sci.* 7:1003 (1963).
11. Gent, A. N., and J. F. McGrath. *J. Polym. Sci.* Part A. 3:1473 (1965).
12. Gent, A. N., and H. Hirakawa, *J. Polym. Sci.* Part A. 5:157 (1967).
13. Andrews, E. H., D. Barnard, M. Braden, and A. N. Gent, "Ozone Attack on Rubber" in *The Chemistry and Physics of Rubber-Like Substances* Ch. 12. L. Bateman, Ed. (New York: John Wiley & Sons, Inc., 1963), p. 329.
14. Erickson, E. R., R. A. Bernsten, E. L. Hill, and P. Kusy, *Rubber Chem. Technol.* 32:1062 (1959).
15. Edwards, D. C., and E. B. Storey. *Rubber Age* 79:5 (1956).
16. Vacca, G.N., *Rubber Chem. Technol.* 32:1080 (1959).
17. Teppema, J. (to The Goodyear Tire & Rubber Company), U.S. Pat. 1,746,371 (1930).
18. Semon, W. L. (to B.F. Goodrich Co.). U.S. Pat. 1,965,948 (l934).
19. Geer, W. C., and C. W. Bedford, *Ind. Eng. Chem.* 17:383 (1925).
20. Semon, W. L. "History and Use Of Materials Which Improve Aging" in *Chemistry and Technology of Rubber* Ch. XII. C.C. Davis and J. T. Blake, Eds. (New York: Reinhold Publishing Corp., 1937), p. 414.
21. Kearsley, E. P. W., *Rubber Age* 27:649 (1930).
22. Kelly, A., B. S. Taylor, and W. N. Jones. *Rubber Chem. Technol.* 1:106 (1928).
23. Harris, J. O., and W. P. Metzner (to Monsanto Chemical Co.). U.S. Pat. 2,748,100 (1956).
24. Semon, W. L., and A. W. Sloan (to B.F. Goodrich Co.), U.S. Pat. 2,000,039 (1935).
25. Semon, W. L., and A. W. Sloan (to B.F. Goodrich Co.), U.S. Pat. 2,000,040 (1935).
26. Shaw, R. F., Z. T. Ossefort, and W. J. Touhey, *Rubber World* 130:636 (1954).

27. Ordnance Corps, Ordnance Spec. MIL-6-12459F C040996-7 (Sept. 24, 1956).
28. Cox, W. L., *Rubber Chem. Technol.* 32:364 (1959).
29. Ambelang, J. C., R. H. Kline, O. M. Lorenz, C. R. Parks, C. W. Wadelin, and J. R. Shelton, *Rubber Chem. Technol.* 36:1497 (1963).
30. Thelin, J. H., and A. R. Davis, *Rubber Age* 86:81 (1959).
31. Spacht, R. B. (to The Goodyear Tire & Rubber Company), U.S. Pat. 3,432,460 (1969).
32. Milner, P. W., and H. Widmer, paper presented at the 5th Int. Rubber Symposium (1975), Paper C15, Gottwaldov.
33. Pospíšil, J. "Aromatic Amine Antidegradants" in *Developments in Polymer Stabilisation* 7:Ch. 1. G. Scott, Ed. (London: Elsevier Applied Science Publishers, 1984), p. 1.
34. Razumovskii, S. D., and G. E. Zaikov, "Degradation and Protection Of Polymeric Materials in Ozone" in *Developments in Polymer Stabilisation* 6:Ch. 6. G. Scott, Ed. (London: Elsevier Applied Science Publishers, 1983), p. 239.
35. Ambeland, J. C., and B. W. Habeck, *Rubber World* 141:96 (1959).
36. Cox. W. L. (to Universal Oil Products). U.S. Pat. 3,304,284 (1967).
37. Walker, L. A., and J. J. Leucken. *Elastomerics* May, 1980, p. 36.
38. Miller, D. E., R. W. Dessent, and J. A. Kuczkowski, *Rubber World* 193:31 (1985).
39. Pinazzi, C. P., and M. Billuart. *Rubber Chem. Technol.* 28:438 (1955).
40. Watanabe, T., Y. Wada, S. Ikeda, M. Iwata. *J. Soc. Rubber Ind. Japan* 41:605 (1968).
41. Watanabe, T., Y. Wada, N. Ikeda, M. Iwata, *J. Soc. Rubber Ind. Japan* 41:613 (1968).
42. Watanabe, T., Y. Sakuramoto, Y. Wada, and I. Ishida. *J. Soc. Rubber Ind. Japan* 41:657 (1968).
43. Watanabe, T., Y. Sakuramoto, Y. Wada, and M. Iwata. *J. Soc. Rubber Ind. Japan* 41:664 (1968).
44. Layer, R. W. "Non-staining Antioxidants," in *Developments in Polymer Stabilisation* 4:Ch. 5. G. Scott, Ed. (London: Elsevier Applied Science Publishers, 1981), p. 135.
45. Brück, D., H. Königshofen, and L. Ruetz. *Rubber Chem. Technol.* 58:728 (1985).
46. McCool, J. C., *Rubber Chem. Technol.* 37:583 (1964).
47. Layer, R. W., *Rubber Chem. Technol.* 39:1584 (1966).
48. Lake, G. J., *Rubber Chem. Technol.* 43:1230 (1970).
49. Andries, J. C., D. B. Ross, and H. E. Diem, *Rubber Chem. Technol.* 48:41 (1975).
50. Andries, J. C., C. K. Rhee, R. W. Smith, D. B. Ross, and H. E. Diem, *Rubber Chem. Technol.* 52:823 (1979).
51. Lattimer, R. P., E. R. Hooser, H. E. Diem, R. W. Layer, and C. K. Rhee. *Rubber Chem. Technol.* 53:1170 (1980).
52. Lattimer, R. P., E. R. Hooser, R. W. Layer, and C. K. Rhee, *Rubber Chem. Technol.* 56:431 (1983).
53. Murray, R. W., in *Polymer Stabilisation*, W. L. Hawkins, Ed. (New York: Wiley-Interscience, 1972), p. 215.
54. Amsden, C. S. *Trans. Inst. Rub. Ind.* 13:T91 (1966).
55. Lorenz, O., and C. R. Parks. *Rubber Chem. Technol.* 36:194 (1963).
56. Lorenz, O., and C. R. Parks, *Rubber Chem. Technol.* 36:201 (1963).
57. Loan, L. D., R. W. Murray, and P. R. Story, *J. Inst. Rubber Ind.* 2:73 (1968).
58. Lorenz, O., F. Haulena, and B. Braun, *Kautschuk & Gummi* 38:255 (1985).
59. Lattimer, R. P., R. W. Layer, and C. K. Rhee, *Rubber Chem. Technol.* 57:1023 (1984).
60. Razumovskü, S. D., and L. S. Batashova, *Rubber Chem. Technol.* 43:1340 (1970).
61. Lattimer, R. P., J. Gianelos, H. E. Diem, R. W. Layer, and C. K. Rhee, *Rubber Chem. Technol.* 59:263 (1986).

CHAPTER 7

Assessment of Ozone Toxicity with Animal Models

Brenda E. Barry

INTRODUCTION

Ozone is the major oxidant present in the photochemical smog of many metropolitan areas. To understand the effects of this air pollutant on the large populations who breathe air containing ozone, animal models are essential. Data gathered from animals exposed acutely and chronically to ozone provide a quantitative base for extrapolation of the effects measured in animals to predictable estimates of the effects in man. Important questions to be addressed with such studies include the structural, physiological, and biochemical changes produced by ozone, especially in the lungs. Progression and reversibility of these changes, and the relevance of age and overall health status to ozone sensitivity are also important questions. Epidemiological studies of exposed human populations alone are inadequate to address these issues. Animal models provide essential information to evaluate the effects of long-term ozone inhalation because such exposures cannot be done ethically with human subjects. Another major advantage of animal models is that exposure conditions can be carefully monitored and controlled throughout an experiment, and the effects of a single pollutant or a specific combination can be studied. In this chapter, the chemistry of ozone will be briefly reviewed as a basis for understanding its mechanisms of toxicity in humans and animals. Different experimental approaches to evaluate ozone toxicity will then be reviewed. Methods to be discussed include biochemistry, physiology, qualitative and quantitative evaluation of the tissue alterations produced by ozone, as well as in vitro assay systems.

MECHANISMS OF OZONE TOXICITY

Ozone has two unpaired electrons in its outer shells which make this molecule an excellent oxidant.[1] Although the mechanisms by which ozone damages biological tissues are not yet clearly delineated, several reactive processes have been recognized. The first is that ozone can oxidize polyunsaturated fatty acids (PUFA) in cell membranes by a reaction termed the Criegee mechanism.[2] This mechanism is currently regarded as the major mode of ozone reactivity in biological systems. It results in the formation of ozonides, which can subsequently decompose to form free radicals, hydroperoxides, and aldehydes. Pryor et al.[3] have reported that free radicals are indeed detectable when ozone reacts with a simple model of a PUFA-containing membrane. Free radicals are very reactive molecular species which have an odd number of electrons. They can interact with a variety of cell components, including DNA, lipids, and proteins, and produce cell damage.[4] Free radicals can initiate lipid peroxidation in cell membranes, and experiments have shown that indices of lipid peroxidation, such as malondialdehdye production, increase after ozone exposure.[5] Additional evidence for the oxidative action of free radicals in ozone injury is that antioxidants, such as vitamin E and free radical scavengers, are protective against ozone injury in animal models.[6] Thus, although free radicals are not directly produced by the Criegee reaction, their formation appears to be an important contributor to ozone damage.

Another reactive process which can result from ozone exposure is the direct oxidation of amino acids in tissue proteins and small peptides.[7] Enzyme susceptibility to ozone is variable and depends upon amino acid composition. Several enzymes, including lysozyme, beta-glucuronidase, and acid phosphatase, can be inactivated by exposure to ozone. In comparison, urease is very resistant.

Because cell membranes contain both proteins and PUFA, it is difficult to determine which component is more critically damaged and it is most likely a combination of the two. The direct effects of ozone and the formation of free radicals are both potentially injurious to membranes and other cell components. The important result is that cellular regulation of electrolytes can be lost, enzymes can be inactivated, and mitochondrial metabolism can be altered. Normal function is disrupted and cell death may occur.[8]

DESIGN CONSIDERATIONS FOR OZONE EXPERIMENTS

Several important factors must be considered in the design of ozone experiments. The first is the exposure conditions and how the ozone concentration will be monitored. Chambers which are specifically designed for the size and type of animal chosen for the ozone exposure are ideal. The ozone concentration should be calibrated by one of several available techniques and then continuously monitored during the exposure period.[9]

A second important factor is the concentration of ozone that will be used. When ozone is inhaled, it is absorbed all along the respiratory tract. Because it is a reactive molecule, its concentration decreases as it penetrates into the airways and parenchyma, or gas exchange region, of the lung. As the concentration of inhaled ozone is increased, there are concomitant increases in the depth of penetration into the lungs and the degree of injury produced. Increases in the rate and depth of respiration of the animals similarly affect the distribution of ozone into the lungs.

The species of animal used for ozone experiments is a most important factor because the extent of injury produced by ozone varies among species. In a 1957 study, Stokinger and his colleagues[10] exposed several different species to 1 ppm ozone for up to 62 weeks. Morphologic examination of lung tissue was used to evaluate chronic lung injury produced by ozone. They reported that there was a range of susceptibility to ozone exposure which increased progressively in the order dogs, mice, hamsters, rats, and guinea pigs.

As illustrated by the Stokinger study, species differences in lung structure, especially airway branching patterns, can influence responses to ozone inhalation. Complex dosimetry models of ozone concentrations within the lungs of a specific species require additional airway parameters be considered. A model described by Miller[11] included radial and axial diffusion of ozone in the airways of each species modeled and the reactivity of ozone with substances that line the airways, such as mucus and surfactant.

Duration of the experiment must also be determined. Exposures can be acute, subchronic, or chronic. Hours per day of exposure must also be defined. As previously mentioned, a major advantage of animal models is that chronic studies, which cannot be done ethically with human subjects, can be conducted with animal models.

Age of the animal at the beginning of the experiment should also be considered. For instance, Stephens et al.[12] have reported that newborn rats are resistant to the effects of ozone until they are weaned and that this factor influences their susceptibility to oxidant injury.

EXPERIMENTAL APPROACHES TO EVALUATE OZONE TOXICITY

Biochemical Measurements

Biochemical changes produced by ozone exposure can be measured by using tissue homogenates and specific tissue fractions. These types of analyses are most often done with lung tissue because it is the organ system most directly affected by ozone. Biochemical assays of lung tissue have shown that ozone exposure can alter the levels of enzymes which are part of intracellular defense systems against injury. These enzymes are important scavengers of free radicals, lipid peroxides, and hydrogen peroxide. Chow et al.[13] reported significant increases in the activities of glutathione peroxidase, glutathione reductase, and associated enzymes after adult rats were exposed to 0.8 ppm ozone for only 3 days. Other antioxidant enzymes within the cell cytoplasm, such as the superoxide dismutases and catalase, also contribute as scavengers of free radicals. Vitamin E (tocopherols) and Vitamin C (ascorbate), lipid soluble scavengers of free radicals, can influence the degree of injury produced by ozone. Animals that are fed diets deficient in these vitamins are more susceptible to ozone injury than those fed normal diets.[14,15]

Additional biochemical analyses for ozone toxicity include increased synthesis of collagens and collagen precursors. This is because fibrotic changes in the lungs have been associated with ozone exposure.[16,17] Formation of malondialdehyde, a product of peroxidation of a PUFA that contains three or more double bonds, is also used as an indicator of lipid peroxidation caused by ozone exposure.[1]

Physiological Measurements

Ozone inhalation can result in altered pulmonary function measurements in humans and animals. Plethysmographs specifically designed for small laboratory animals can be used to measure tidal volume, respiratory frequency, and peak flow rates. Pressure-volume curves can also be determined and calculations can be made for dynamic compliance, total lung volumes, and other volume measurements.

The effects of ozone inhalation have been measured in a variety of species, including guinea pigs, cats, dogs, and monkeys, as reviewed by Mauderly.[18] The results of acute ozone exposure studies showed that there were comparable responses between humans and other species, including increases in respiratory frequency and decreases in tidal volume. Comparison of the effects of 2 hours exposure to approximately 0.5 ppm ozone in humans, guinea pigs, and cats indicated similar and progressive increases in airway constriction. Six-week exposure experiments by Raub et al.[19] showed that adult rats exposed to 0.25 ppm ozone had significant increases in end-expiratory volumes. Newborn rats exposed under the same conditions had increases in inspiratory capacity, vital capacity, and total lung capacity, suggesting that ambient levels of ozone may affect lung growth and development in the rat. Pulmonary function tests on monkeys exposed to 0.5 or 0.8 ppm ozone for up to 90 days indicated a general trend of increased quasistatic compliance, but overall pulmonary function values did not statistically differ before and after exposure.[20]

These reports illustrate the similarity between the acute responses of humans and animals to the effects of ozone. They also suggest that chronic exposures in animals may help predict the adverse effects of ozone in humans exposed for prolonged periods.

Qualitative Evaluation of Lung Tissue

As previously discussed, ozone is absorbed all along the respiratory tract. It is not clearly understood how much ozone is removed by the upper airways which include the nasal cavity, larynx, and pharynx. Miller et al.[11] estimate that 50% of the inhaled ozone is taken up by this region. Morphologic examination of the trachea, bronchi, and larger bronchioles indicate that the ciliated cells of these airways are most affected. These cells are an important component of the respiratory defense system because, by the motion of their cilia, they serve to remove inhaled particulates, bacteria, and other debris from the lungs and keep them sterile. The changes in these cells produced by ozone include shortening of the cilia, the presence of cell blebs, and loss of cells.[21]

Although there are distinct differences among species in airway branching patterns and distances from the trachea to the gas exchange region, ozone inhalation most often produces a lesion at the junction between the end of the conducting airways and the gas exchange region.[10] This area is termed the centriacinar region. In small animals, such as rats, mice, and guinea pigs, the last segment of the conducting airways, the terminal bronchioles, branch very quickly into alveolar, or gas exchange, tissue. In larger animals, such as dogs, primates, and also in man, the transition between the last segment of conducting airways and alveolar tissue is more gradual. It consists of a combination of airway lining cells and alveolar cells, and this structure is more accurately termed a respiratory bronchiole. Regardless of species differences, this junctional area, or centriacinar region, is a primary site of injury produced by ozone inhalation.

Two cell types line the majority of the terminal bronchiolar surface. First, there are ciliated cells, and both their function and the damage that ozone produces in the cells have been described above. The second cell type is the nonciliated bronchiolar epithelial cell, which is often called more simply the Clara cell. The function of this cell type is not well understood, but it is believed that it secretes components of the material lining the terminal bronchioles, metabolizes foreign compounds that reach this region of the lungs, and is the progenitor cell for renewal of the epithelium.[22] Ozone inhalation produces alterations in the shape of these cells, including decreases in the size of the distinctive bulbous projection of these cells into the airway lumen.

Different cell types line the adjacent alveolar tissue in the centriacinar region (Figure 1). Type I epithelial cells are relatively large, flat, very thin cells which line the majority of the alveolar surface. Because they have a large exposed membrane surface, they are very susceptible to the oxidative effects of the ozone molecule. Examination of the centriacinar region of rats after only 2 hours of exposure to 0.2 ppm ozone indicated that these cells were swollen, had membrane breaks, and in some areas had sloughed off the underlying alveolar tissue.[23] Type II epithelial cells, which are more compact, cuboidal cells, line almost all of the remaining alveolar surface and produce surfactant, a material which lines the alveolar air surface. These cells are more resistant to the effects of ozone and their numbers increase after ozone exposure because they serve as progenitor cells for replacement of the damaged type I epithelial cells (Figure 2). Ozone inhalation also increases the numbers of macrophages in the airspaces of the small airways and the adjacent alveolar tissue. Macrophages are another important

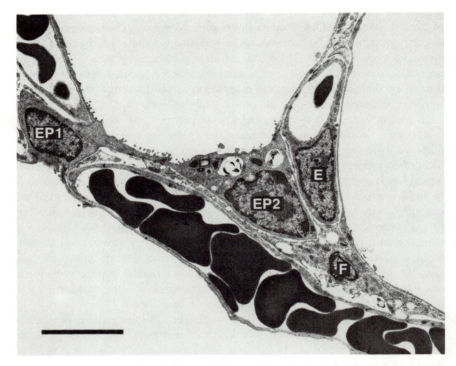

Figure 1.   Alveolar septa from a 12 week old control rat. A cuboidal epithelial type II (EP2) and an attenuated type I epithelial cell line the alveolar surface. A fibroblast (F) in the interstitium and an endothelial cell (E) are also present. Bar = 10 microns. From reference 24.

component of the respiratory defense system and they serve to remove debris and have antibacterial mechanisms.

It is evident from the preceding discussion that ozone does not produce uniform injury in the lung. As the concentration of inhaled ozone approaches the National Ambient Air Quality Standard (NAAQS) of 0.12 ppm ozone, injury produced by these lower concentrations of ozone becomes more difficult to detect. This is because biochemical and physiological measurements generally represent alterations produced in the lungs as a whole. Selective injury to specific parts of the lungs, especially at lower ozone concentrations, may be diluted out by averaging in all parts of the lungs in these measurements. From morphologic studies, it is known that the centriacinar region is an area of focal injury, but alterations produced here by concentrations of ozone near the NAAQS have been reported as minimal to slight.[15]

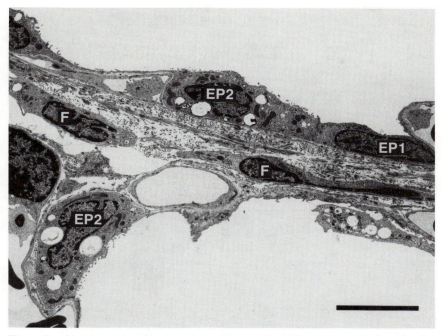

Figure 2. Alveolar septum from an adult rat exposed to 0.25 ppm ozone for 6 weeks. Two type II epithelial cells (EP2) and a type I epithelial cell line the majority of this alveolar septum. A fibroblast (F), an interstitial macrophage (M), and increased amounts of collagen are present in the interstitium. Bar = 10 microns. From reference 24.

Quantitative Evaluation of Lung Tissue

Quantitative methods, based upon morphometric principles, have been developed to selectively evaluate changes produced by ozone in the centriacinar region.[24,25] These techniques incorporate geometric and statistical methods to quantify selected cell parameters from micrographs of lung tissue. To apply these methods, terminal bronchioles and their associated alveoli were isolated from the lungs of rats after exposure to 0.25 ppm ozone for 6 weeks. To evaluate the effects of animal age and extent of injury, rats were either 1 day old (juveniles) or 6 weeks old (adults) at the beginning of the exposure. Results of changes produced by ozone in the alveolar epithelial cells of the centriacinar region are illustrated in Figure 3. These results show that this exposure produced statistically significant alterations in the number, size, and shape of these epithelial cells, which represents chronic epithelial injury. Similar analyses of 6-week-old animals exposed to 0.12 ppm ozone showed smaller, but significant changes in the type

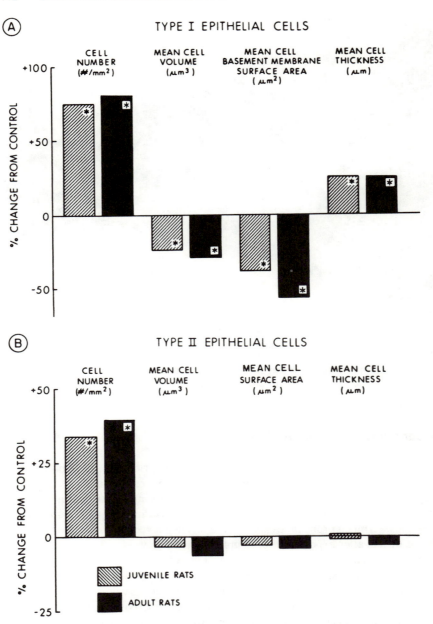

Figure 3.   Epithelium in the centriacinar region. The changes in the morphometric characteristics of type I and type II epithelial cells of juvenile and adult rats expressed as percent change from control values following exposure to 0.25 ppm ozone for 6 weeks. *p < .05. From reference 24.

I epithelial cells, suggesting a relatively linear relationship between inhaled ozone concentration and alterations in the epithelial tissue. Changes in the ciliated and nonciliated (Clara) cells from terminal brochioles of the same groups of rats were also quantified. The results indicated that 6 weeks of exposure to 0.25 ppm ozone also produced changes in numbers and characteristics of these cells.

The reasons for the particular susceptibility of the centriacinar region are not well understood, but several are proposed. The first is the absence of mucus-producing goblet cells in the terminal and respiratory bronchioles. Mucus is secreted from the goblet cells onto the surface of the larger upper airways, and serves to trap inhaled particles. Mucus can react with ozone, protecting the underlying epithelium from its oxidative effects.[11] If any mucus is present in the terminal bronchioles, the layer is presumably very thin and cannot afford adequate protection to the underlying epithelial cells. A second reason is that mass movement of gases is thought to end at the terminal bronchioles, and that diffusion alone occurs in adjacent alveoli. Because of the reactive nature of ozone, the alveolar tissue in the centriacinar region would be the primary target of injury. The dosimetry model proposed by Miller[11] also indicated that the centriacinar region receives the highest concentration of ozone.

The centriacinar region has significant functional importance because it is the intersection between the conductive and respiratory portions of the lungs. It is the site of several pathologic lesions, including loss of alveolar tissue, as seen in centrilobular emphysema, and fibrotic changes produced by oxidant gases, such as ozone. Such structural changes may affect gas exchange in this region and alter normal distribution of gases to more distal portions of the lungs.

In Vitro Assays

Cells can be isolated from the lungs of animals to evaluate how ozone exposure alters a specific cell population. These cells can be isolated following an in vivo ozone exposure, or first isolated from the animals and then exposed in vitro. These assays are often done with alveolar macrophages because these cells can easily be washed out of the lungs with saline solutions. Amoruso and colleagues[26] reported that the production of superoxide anions, a bacterial defense mechanism present in alveolar macrophages, was decreased after ozone exposure. Ozone exposure can also depress the activity of intracellular enzymes, such as acid hydrolases,[27] and decrease the mobility of these alveolar macrophages.[28] These functions are important for the respiratory defense roles that these cells perform in the lungs. Compromise of these functions may increase the risk of host infection.

Haagsman and his associates[29] isolated type II epithelial cells from the lungs of rats, and exposed the cells in vitro to ozone. They reported that the exposure impaired the synthesis of surfactant lipids. These lipids are essential for reducing surface tension at the air-tissue interface in the alveolar region.

SUMMARY

Man is not a large rat, but animal models can provide important data for estimates of the effects of inhaled toxicants in man. As discussed in this chapter, there are a variety of experimental approaches to evaluate the toxicity of ozone. By interpretation and integration of this information, reasonable predictions of the health effects of ozone in man can be made.

REFERENCES

1. Mustafa, M. G., and D. F. Tierney. "Biochemical and Metabolic Changes in the Lungs with Oxygen, Ozone, and Nitrogen Dioxide," *Am. Rev. Resp. Dis.* 118:1061–1090 (1978).
2. Criegee, R. "The Ozonization of Unsaturated Compounds," *Rec. Chem. Progr.* 18:111–120 (1957).
3. Pryor, W. A., M. M. Dooley, and D. F. Church. "Mechanisms for the Reaction of Ozone with Biological Molecules, the Source of the Toxic Effects of Ozone," in *Biomedical Effects of Ozone and Photochemical Oxidants*, S. D. Lee, M. G. Mustafa, and M. A. Mehlman, Eds. (Princeton, N.J.: Princeton Scientific, 1983), pp. 7–19.
4. Freeman, B. A., and J. D. Crapo. "Free Radicals and Tissue Injury," *Lab. Invest.* 47:412–426 (1981).
5. Chow, C. K., and A. L. Tappel. "An Enzymatic Protective Mechanism Against Lipid Peroxidation Damage in Lungs of Ozone-Exposed Rats," *Lipids* 7:518–524 (1972).
6. Chow, C. K. "Influence of Dietary Vitamin E on Susceptibility to Ozone Exposure," *Adv. Mod. Environ. Toxicol.* 5:75–93 (1982).
7. Mudd, J. B., and B. A. Freeman. "Reaction of Ozone with Biological Membranes," in *Biomedical Effects of Environmental Pollutants*, S. D. Lee, Ed. (Ann Arbor, Michigan: Ann Arbor Science Publishers, Inc., 1977), pp. 97–133.
8. Menzel, D. B. "Ozone: An Overview of Its Toxicity in Man and Animals," *Adv. Mod. Environ. Toxicol.* 5:183–204 (1982).
9. "Air Quality Criteria for Ozone and Other Photochemical Oxidants," U.S. EPA-Report 600/8-84-020aF, Vol. 1 (1986).
10. Stokinger, H. E., W. D. Wagner, and O. J. Dobrogorski. "Ozone Toxicity Studies. III. Chronic Injury to Lungs of Animals Following Exposure at Low Levels," *Arch. Indust. Health* 16:514–522 (1957).
11. Miller, F. J. "Biomathematical Modelling Applications in the Evaluation of Ozone Toxicity," in *Assessing Toxic Effects of Environmental Pollutants*, Mudd, J. B., and S. D. Lee, Eds. (Ann Arbor, Michigan: Ann Arbor Science Publishers, Inc., 1979), pp.263–286.
12. Stephens, R. J., M. F. Sloan, D. G. Groth, D. S. Negi, and K. D. Lunan. "Cytologic Response of Postnatal Rats Lungs to O_3 or NO_2 Exposure," *Am. J. Path.* 93:183–200 (1978).
13. Chow, C. K., M. Z. Hussain, C. E. Cross, D. L. Dungworth, and M. G. Mustafa. "Effect of Low Levels of Ozone on Rat Lungs. I. Biochemical Responses During Recovery and Reexposure," *Exptl. Molec. Path.* 25:182–188 (1976).

14. Sato, S., M. Kawakami, S. Maeda, and T. Takashima. "Scanning Electron Microscopy of the Lungs of Vitamin E-Deficient Rats Exposed to a Low Concentration of Ozone," *Am. Rev. Resp. Dis.* 113:809–821 (1976).

15. Plopper, C. G., D. L. Dungworth, W. S. Tyler, and C. K. Chow. "Pulmonary Alterations in Rats Exposed to 0.2 and 0.1 ppm Ozone: a Correlated Morphological and Biochemical Study," *Arch. Environ. Health* 34:390–395 (1979).

16. Last, J. A., D. B. Greenberg, and W. A. Castleman. "Ozone-Induced Alterations in Collagen Metabolism in Rat Lungs," *Toxicol. Appl. Pharm.* 51:247–258 (1979).

17. Hussain, Z. M., C. E. Cross, M. M. Mustafa, and R. S. Bhatnagar. "Hydroxyproline Contents and Prolyl Hydroxylase Activities in the Lungs of Rats Exposed to Low Levels of Ozone," *Life Sci.* 18:897–904 (1976).

18. Mauderly, J. L. "Respiratory Function Responses of Animals and Man to Oxidant Gases and to Pulmonary Emphysema," in *Fundamentals of Extrapolation Modeling of Inhaled Toxicants*, F. J. Miller and D. B. Menzel, Eds. (Washington, D.C.: Hemisphere Publishing Corp., 1984), pp. 165–181.

19. Raub, J. A., F. J. Miller, and J. A. Graham. "Effects of Low level Ozone Exposure on Pulmonary Function in Adult and Neonatal Rats," *Adv. Modern Environ. Toxicol.* 5:363–367 (1982).

20. Eustis, S. L., L. W. Schwartz, P. C. Kosch, and D. L. Dungworth. "Chronic Bronchiolitis in Nonhuman Primates After Prolonged Ozone Exposure," *Am. J. Path.* 105:121–137 (1981).

21. Schwartz, L. W., D. L. Dungworth, M. G. Mustafa, B. K. Tarkington, and W. S. Tyler. "Pulmonary Responses of Rats to Ambient Levels of Ozone. Effects of a 7-Day Intermittent or Continuous Exposure," *Lab. Invest.* 34:565–578 (1976).

22. Evans, M. J., L. V. Johnson, R. J. Stephens, and G. Freeman. "Renewal of the Terminal Bronchiolar Epithelium in the Rat Following Exposure to NO$_2$ or O$_3$" *Lab. Invest.* 35:246–257 (1976).

23. Stephens, R. J., M. F. Sloan, M. J. Evans, and G. Freeman. "Early Response of Lung to Low Levels of Ozone," *Am. J. Path.* 74:31–58 (1973).

24. Barry, B. E., F. J. Miller, and J. D. Crapo. "Effects of Inhalation of 0.12 and 0.25 Parts Per Million Ozone on the Proximal Alveolar Region of Juvenile and Adult Rats," *Lab. Invest.* 53:692–704 (1985).

25. Barry, B. E., R. R. Mercer, F. J. Miller, and J. D. Crapo. "Effects of Inhalation of 0.25 ppm Ozone on the Terminal Bronchioles of Juvenile and Adult Rats," *Exptl. Lung Res.* 14:225–245 (1988).

26. Amoruso, M. A., G. Witz, and B. D. Goldstein. "Decreased Superoxide Anion Radical Production by Rat Alveolar Macrophages Following Inhalation of Ozone or Nitrogen Dioxide," *Life Sci.* 28:2215–2221 (1981).

27. Hurst, D. J. and D. L. Coffin. "Ozone Effect on Lysosomal Hydrolases of Alveolar Macrophages In Vitro," *Arch. Intern. Med.* 127:1059–1063 (1971).

28. McAllen, S. J., S. P. Chiu, R. F. Phalen, and R. E. Rasmussen. "Effect of In Vivo Ozone Exposure on In Vitro Pulmonary Alveolar Macrophage Mobility," *J. Toxicol. Environ. Health* 7:373–381 (1981).

29. Haagsman, H. P., E. A. J. M. Schuurmans, G. M. Alink, J. J. Batenberg, and L. M. G. vanGolde. "Effects of Ozone on Phospholipid Synthesis by Alveolar Type II Cells Isolated from Adult Rat Lung," *Exptl. Lung Res.* 9:67–84 (1985).

CHAPTER 8

The Respiratory Effects of Low Level Ozone Exposure: Clinical Studies

William F. McDonnell

INTRODUCTION

Ozone, a chemical oxidant and respiratory irritant, is a major component of photochemical smog. The present National Ambient Air Quality Standard (NAAQS) for ozone is a maximum daily one-hour average not to exceed 0.12 part per million (ppm) on more than one occasion per year. Many areas of the United States are not in attainment of this standard, with millions of people routinely exposed to ambient air containing levels of ozone at which acute effects upon the respiratory system have been demonstrated. The purpose of this chapter is to review results of some of the controlled chamber studies of humans exposed to ozone. In doing this, the respiratory effects which are known to occur during and following an acute exposure to ozone will be described. This will be followed by a description of the conditions under which effects can be expected in man, and of the populations in which we expect to observe these effects. The chapter will conclude with discussion of the health consequences of these effects, and with the phenomenon of attenuation of response following repeated exposures.

DESCRIPTION OF ACUTE EFFECTS

When sufficient ozone is inhaled for one to two hours, a cascade of events is set in motion which takes several days to completely resolve. Ozone first reacts chemically with compounds in the surface liquid of the respiratory tract, producing a number of reactive compounds which, like ozone, can undergo oxidative reactions with other substances. Either ozone, or these reactive substances, or both, then react chemically with the epithelial cells lining the airways and alveoli, resulting in death and damage to some of these cells. These damaged cells release a number of biologically active substances into the airway and surrounding tissue.

In humans we observe a number of responses of the respiratory system to ozone, the other oxidative molecules, and/or these released mediators. These responses include both changes in lung function and symptoms which appear early during exposure and which generally resolve within hours to a day, and effects that are less obvious and take longer to resolve.

The symptoms most commonly seen following ozone exposure include the induction of cough, pain upon deep inspiration, and shortness of breath.[1-3] Accompanying these symptoms are changes in a number of respiratory function measures.[1-5] Exposed individuals experience decrements in forced vital capacity (FVC), forced expiratory volume in one second (FEV_1), and other forced expiratory flow variables including peak expiratory flow (PEF) and flow rate between 25% and 75% of FVC (FEF_{25-75}). The decrease in FVC is due to limitation in the ability of the subject to inhale maximally with a reduction in total lung capacity (TLC) and little or no change in residual volume.[2,3,6] This inability to inhale is probably mediated by neural receptors in the airways and possibly the walls of the alveoli, and is not the result of mechanical changes in the lung.[7] The decrements in forced expiratory flow are at least partially accounted for by the inability to inhale maximally, as flow rates are a function of the volume at which they are measured. These decrements may be augmented by airway narrowing which also occurs during ozone exposure,[1,2,6] as indicated by increases in specific airway resistance (SRaw). This, too, is a neurally mediated phenomenon and is the result of contraction of the smooth muscle surrounding the airways.[6] This constriction of the airways is apparently independent of the inability to inhale maximally, as the magnitudes of these two effects are not statistically associated,[1] and bronchoconstriction can be blocked by administration of atropine while the reductions in TLC and FVC cannot.[6] Exposure to ozone also results in the onset of rapid, shallow breathing,[1,3] which is also neurally mediated. While these may be protective responses of the respiratory system, their immediate effect is to cause discomfort and to limit performance.

More recently, further acute responses to a single exposure have been documented. These include the induction of an inflammatory response in the lung, changes in the barrier function of the epithelium lining the airways and alveoli, and an increase in nonspecific airway responsiveness. Within three hours following exposure, an increase in the number of polymorphonuclear leukocytes (PMN) has been observed in the fluid from bronchoalveolar lavage.[8] This has been accompanied by an increase in the concentration of several arachadonic acid metabolites, presumably mediators from the damaged epithelial cells or from the

PMNs themselves. Coincident with this has been an increase in the responsiveness of the airway to inhalation of histamine or acetyl, beta-methyl, choline.[8,9] Two hours following exposure, the airways and alveoli have been reported to be more permeable to macromolecules as measured by their rate of disappearance from the lung.[10] Because of the recent nature of these observations, little is known about the time course or significance of these effects. Furthermore, in studies describing these changes, levels of ozone have usually been at or above 0.40 ppm. Consequently, little information about effects at lower levels is available.

CONDITIONS UNDER WHICH ACUTE EFFECTS OCCUR

The magnitudes of responses to ozone exposure have been demonstrated to be functions of ozone concentration, minute ventilation during exposure, and duration of exposure.[4,11,12] The relative importance of each of these variables, however, is not equal, and the complex relationship between response and these variables has not been elucidated. It does appear that ozone concentration is more important in determining magnitude of effect than are the other two variables, and the concentration-response relationships for some acute effects have been described at fixed levels of ventilation and duration of exposure.

In order to describe the relationship of response to ozone concentration, a study was performed to measure the effects of six different levels of ozone exposure upon subjects who were alternating 15 minute periods of rest with 15 minute periods of heavy treadmill exercise for two hours of an ozone exposure.[1] The level of exercise chosen is one that can be maintained for this duration by healthy young men, and would be exceeded only in circumstances such as competitive cycling, soccer, or long-distance running. As such, this represents a near worst-case situation for two-hour exposures. This duration was chosen as representative of the time during which peak levels of ozone existed in the ambient air. We have since recognized that peaks are often superimposed upon longer periods during which ozone is present at lower levels, and that under certain conditions peak ozone levels are sustained for longer periods. This will be discussed in a later section.

In this particular study, 135 normal healthy men, age 18–30 years, without a history of allergy, asthma, or hay fever were selected for participation. Subjects were divided into six groups, and each group was exposed to one of the following ozone concentrations: 0.0, 0.12, 0.18, 0.24, 0.30, and 0.40 ppm. Immediately prior to exposure, subjects performed forced expiratory spirometry measures of FVC, FEV_1, and FEF_{25-75}, followed by measurement by SRaw. They then completed symptom questionnaires inquiring about the presence of cough, shortness of breath, pain on deep inspiration, and several sham symptoms. Symptoms were rated as none, mild, moderate, and severe, and were scored as 0, 1, 2, and 3, respectively. The first two hours of exposure consisted of alternating 15 minute periods of rest and treadmill exercise, followed by 30 minutes during which time final measures of lung function and symptoms were made. Measurements of minute ventilation (V_E), respiratory rate (f), and tidal volume (V_T) were made during the twelfth minute of each exercise period. The effects of exposure on f and V_T were assessed by comparing the values for the first and last exercise periods, while differences between the baseline and final measurements were used for the other variables.

All subjects had normal baseline lung function. The mean treadmill speed was between 7.6 and 7.9 kilometers per hour and the mean grade of incline was between 11.3 and 12.1% for all groups. The resultant values for V_E, heart rate, and oxygen consumption were 65.6 liters min⁻¹, 161 beats per minute, and 2.64 liters min⁻¹, respectively. The differences in these variables among the six groups were small and of no significance.

The results indicate that there were small average decrements in FVC (3%), FEV_1 (5%), and FEF^{25-75} (7%) for the group exposed to 0.12 ppm which were significantly different from the group exposed to 0.0 ppm (control group). These decrements were progressively larger in the groups exposed to higher concentrations of ozone (Figure 1). The changes in SRaw, f, and V_T of the 0.12 and 0.18 ppm groups were not different from the changes of the control group (Figure 2). At 0.24 ppm and above, however, significant increases in SRaw and f and a significant decrease in V_T were observed. The symptom cough was significantly increased in the group exposed to 0.12 ppm (43% with mild cough, and 19% with moderate cough) compared to the control group, with larger increases at higher concentrations of ozone. There were no significant increases in the symptom pain on deep inspiration or shortness of breath for the 0.12 or 0.18 ppm groups compared to the control group. There were significant increases in both of these symptoms at the levels 0.24 ppm and above (Figure 3). Similar concentration-response relationships have been observed in other studies, with some differences among the studies in the magnitude of response at some ozone concentrations.[2,13]

INDIVIDUAL VARIABILITY IN RESPONSE

The above results are mean changes for each group as a whole. During the course of the study we were impressed that individuals were responding very differently to similar exposures. This is further illustrated in Figure 4, which contains frequency histograms of the individual changes in FEV_1 and SRaw at each of the six concentrations of ozone. For the control group, the majority of subjects experienced changes in FEV_1 of between ± 2%, with one subject having a decrement of between 5 and 10%. At 0.12 ppm, the distribution was shifted to the right, as would be expected from the greater mean response. At this level, of 22 subjects; there were still subjects with no significant decrease in FEV_1; two subjects experienced 10% to 15% decrements; and one subject had a 15%

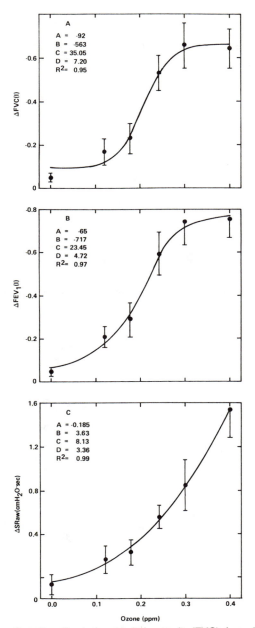

Figure 1. Changes from baseline in forced vital capacity (FVC), forced expiratory volume in 1 second (FEV$_1$), and specific airway resistance (SRaw) as functions of ozone concentration. Mean data are fitted by logistic functions of the form $Y = A + B / [1 + \exp(-CX + D)]$, where X = ozone concentration in ppm, Y = change from baseline for FVC and FEV$_1$ in mL, and SRaw in cm H$_2$O sec.

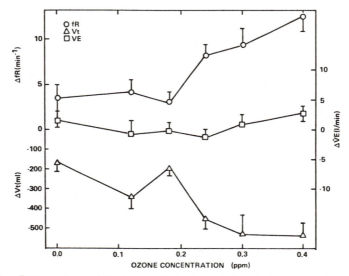

Figure 2. Percent change from baseline for minute ventilation (Vᴇ), tidal volume (Vᴛ), and respiratory rate (fʀ).

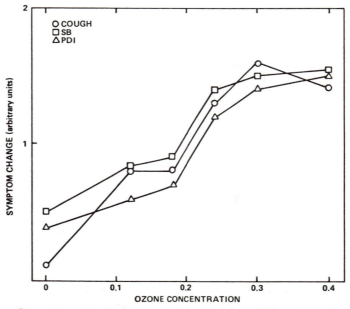

Figure 3. Change from baseline in symptom severity for cough, shortness of breath (SB), and pain on deep inspiration (PDI). Units of severity before and after exposure were none = 1, mild = 2, moderate = 3, and severe = 4.

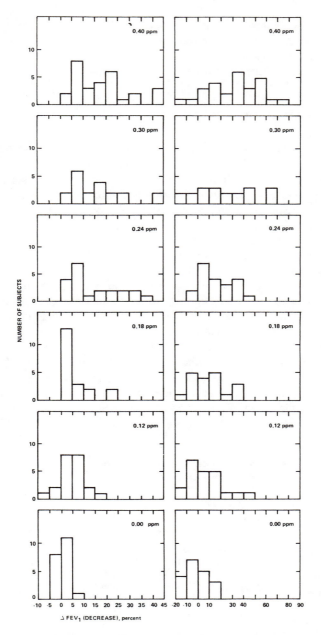

Figure 4. Frequency distributions of response (percent change from baseline) in specific airway resistance (SRaw) and forced expiratory volume in 1 second (FEV$_1$) for individuals exposed to 6 concentrations of ozone. One individual with a 260% increase in SRaw exposed to 0.4 ppm ozone is not graphed.

to 20% decrease. As the concentration increased, the range of responses increased, with some subjects experiencing essentially no change in FEV_1, and others experiencing decrements of greater than 40%. This pattern was also seen in the SRaw data and for all other variables.

REPRODUCIBILITY OF RESPONSES

The reasons for the observed differences in magnitude of response are not clear. In order to determine whether these differences represented true differences in the responsiveness of individuals to ozone, or whether they had some other explanation, we undertook a study of the reproducibility of the responses.[14] Several subjects from the previous study were asked to return for reexposure under identical conditions, and new subjects were recruited to undergo either two or three exposures. A total of 32 subjects had at least two exposures, with duration between exposures varying from 3 weeks to 1 year. Figure 5 indicates that, in general, subjects with a large decrement in FEV_1 following the first exposure experienced a large response on subsequent exposure. Table 1 presents the intraclass correlation coefficient for each of the variables at each level of ozone. This intraclass correlation coefficient measures the fit of data to the line of identity in the same way that a standard correlation coefficient measures the fit of data to best-fitting least square regression line. In general, these data indicate that the spirometric variables and the symptom cough are quite reproducible. The reason for the lower values at 0.12 and the unstable values at 0.30 ppm is that by chance the subjects exposed to these levels had a fairly narrow range in response. Had a larger number of subjects been exposed at each level, we probably would have observed a larger range and better reproducibility. The variables

Table 1.   Intraclass Correlation Coefficients for Changes Induced by Repeat Exposure for Five Concentrations of Ozone.

	R				
O_3(ppm)	0.12 (n = 8)	0.18 (n = 8)	0.24 (n = 5)	0.30 (n = 5)	0.40 (n = 6)
FVC(L)	0.26	0.97[c]	0.88[b]	0.88[b]	0.87[b]
FEV_1(L)	0.57	0.94[c]	0.88[c]	0.56	0.91[c]
FEF_{25-75}(L/sec)	0.45	0.90[c]	0.83	0.35	0.85[b]
SRaw(cm/sec)	0.38	0.62	0.96[c]	0.72	0.48
Cough	0.64[a]	0.78[b]	0.80	0.70	1.00[c]
S.B.	0.05[a]	0.46	0.71	0.58	0.88[c]

Definition of abbreviations: R = intraclass correlation coefficient; O_3 = ozone concentration in part per million; FVC = forced vital capacity; FEV_1 = forced expiratory volume in one second, FEF_{25-75} = mean expiratory flow rate between 25% and 75% of FVC; SRaw = specific airway resistance; S.B. = shortness of breath.

[a]Number of subjects = 7.

[b]$p < .01$ for R = 0.

[c]$p < .001$ for same.

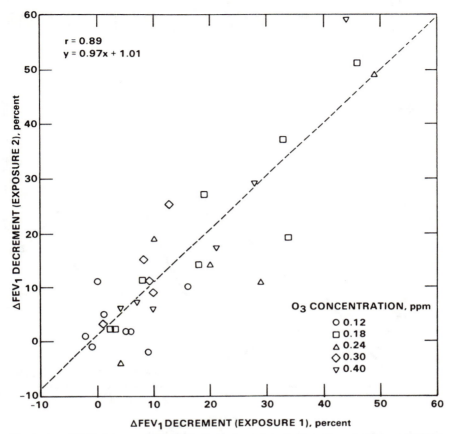

Figure 5. Percent decrease from baseline in forced expiratory volume in 1 second (FEV₁) for 32 subjects exposed to ozone on two occasions.

SRaw and shortness of breath were fairly reproducible with the variables f, V_T, and pain on deep inspiration not being reproducible. These data clearly indicate that some individuals are more responsive to ozone exposure than are others. Similar results for changes in FEV_1 have been observed by others for exposures to 0.4 ppm ozone.[15]

In order to further document these differences, we selected three subjects with different responses and asked them to undergo exposures to several concentrations. The individual concentration-response curves clearly demonstrate these intraindividual differences at more than a single concentration of ozone (Figure 6).

We conclude that this spectrum of responses represents true differences in intrinsic responsiveness to ozone in this otherwise homogeneous group of healthy young men. We have attempted without success to find differences among these subjects in variables, such as baseline nonspecific airway reactivity, to account

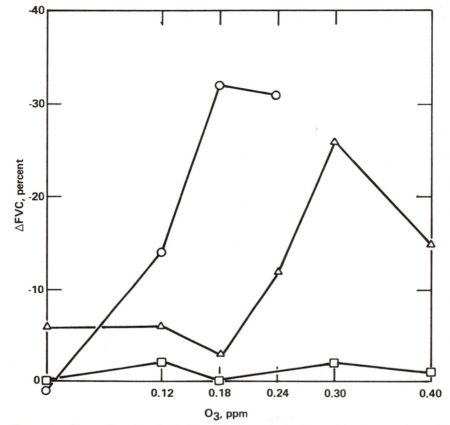

Figure 6.   Change from baseline in forced vital capacity for three subjects exposed to multiple concentrations of ozone.

for these differences in ozone responsiveness. One possible explanation that we are currently testing is based on the recent observation that different individuals remove different amounts of inspired ozone in the upper airways, resulting in different doses to the lower airways.[16]

It should also be pointed out that there is variability in all biological systems. It is of considerable interest in this case because individuals are frequently exposed to ambient air containing ozone at concentrations at which these differences become obvious.

COMPARATIVE RESPONSE AMONG POPULATIONS

Frequently concern has been expressed that some subgroups of the general population may be more responsive to ozone or more susceptible to the effects produced by exposure. A number of clinical studies have been performed to test this hypothesis. Subjects with allergic rhinitis have been tested under conditions known to produce effects, and have been found to have responses similar to those without allergic rhinitis.[17] Similar studies of children and adolescents have been performed at low levels of ozone.[18,19] In general the observed changes in lung function are very similar to those of adults. There are, however, differences between the groups, with children reporting no symptoms at concentrations for which adults have previously reported symptoms. It has been questioned whether these differences are real or are the result of the difficulties in quantifying symptoms experienced by children. Studies of women have demonstrated similar or greater effects in women than expected in men.[3,20,21] However, in many of these studies women have exercised under conditions requiring minute ventilations equal to those of men. When corrected for the differences in body size between men and women, the women received larger ozone doses than did the men. In general, it doesn't appear that women are more responsive than men.

Studies of patients with asthma and with chronic obstructive lung disease have been performed in several laboratories.[22-27] In general, the conditions of exposure were not sufficient to produce measurable changes in healthy individuals. No effects were observed in either of these two groups under the conditions of exposure. The lack of response indicates that these groups are probably not an order of magnitude more responsive than healthy young men, but does not rule out the possibility that there may be some differences in response. Furthermore, the majority of patients studied would be considered to have mild disease. Whether more severely impaired patients would have greater responses is not known.

ATTENUATION OF RESPONSE

If individuals are exposed to 0.4 ppm ozone for two hours daily over a period of a week, the magnitude of the symptoms and lung function decrements is increased on the second day of exposure compared to the first day, and on subsequent days, the decrement decreases and disappears altogether by the fourth day, in most instances.[28-31] This phenomenon of attenuation has historically been referred to as "adaptation." Attenuation of the response at a low level of ozone does not decrease the response to a higher concentration,[15] and in the absence of further exposures, the attenuation disappears gradually over the course of one to two weeks.[30,31] The potential health significance of this phenomenon will be discussed in a later section.

EFFECTS OF PROLONGED EXPOSURES

The majority of the laboratory studies reported on have utilized one or two hour exposures. These were intended to mimic what were initially thought to be the prevalent ambient exposure patterns. As mentioned, recent air monitoring data indicate that large numbers of individuals are exposed to low levels of ozone that persist for considerably longer periods than two hours. A recent epidemiological study has suggested that these longer term exposures may produce larger effects than predicted by the magnitude of the highest one hour peak concentration.[32] In order to further explore this possibility, a study has been performed in which 10 healthy young men were exposed to 0.12 ppm ozone and performed moderate exercise for 50 minutes out of every hour for six hours, with an additional 30 minute break after the third exercise period.[33] This was intended to replicate the conditions under which someone doing heavy outdoor work might be exposed. Preliminary results indicate that the effects after six and a half hours were considerably greater than after two hours, and were larger than were seen after two hours in the study of heavily exercising men previously reported on. It is not possible to say what the results of longer duration studies would be during exposures to levels of ozone below 0.12 ppm. This is presently being investigated.

SIGNIFICANCE OF ACUTE EFFECTS

The interpretation of the health significance of the measured acute responses of humans to ozone exposure can be made at two levels. The acute effects can, of themselves, interfere with normal activities by limiting performance and producing annoying symptoms. In extreme cases of high level ozone exposures which are occasionally experienced under ambient conditions, responsive individuals may actually be unable to continue their normal activities due to severe symptoms and lung function decrements. The other question is whether even small changes in lung function and symptoms are indicators of underlying processes such as cellular damage and inflammation which, if experienced repeatedly, may result in chronic lung disease. Interpretation of data from laboratory exposures is further complicated by the recent understanding that ambient exposure to 0.12 ppm ozone for one hour may also include several hours of exposure to somewhat lower levels just prior to and after the peak. Continued exposures of this type over a long period of time may increase the probability of developing permanent lung damage.

Regarding the acute effects as health problems in and of themselves, it was noted that following a two hour exposure to 0.12 ppm ozone with intermittent heavy exercise, 10% of subjects experienced decrements in FEV_1 between 10% and 15%, with 5% of subjects experiencing a 15% to 20% decrement. Approximately 43% of subjects reported mild cough and 19% reported a moderate cough following exposure. In general, decrements in FEV_1 are the result of inability to inspire deeply, and are often associated with discomfort or coughing on attempting such a breath. In my opinion, individuals experiencing a 10% decrement may be aware of it during heavy exercise and may find their performance limited somewhat. Individuals experiencing a 15% decrement will be more aware and have further limitations, while a 5% decrement is probably not noticeable acutely to the majority of individuals, unless it is accompanied by cough or other symptoms. While a mild cough may be well tolerated by most individuals, the presence of a moderate cough may result in significant discomfort on the part of the exposed individual.

A further theoretical concern of the acute effects of ozone exposure exists for asthmatic individuals. While clinical studies have not found asthmatics to be more responsive to ozone than nonasthmatics, two epidemiologic studies have suggested that frequency of asthma attacks is associated with ozone concentration.[34,35] It is conceivable that the increased airway reactivity documented at higher concentration exposures could result in exaggerated responses to allergens. It is also possible that the observed epithelial permeability increases could allow greater access of inhaled antigen to mast cells.

The question of how to interpret acute effects as indicators of the progression of irreversible lung damage is difficult. The few studies of chronic exposure in animals suggest that recurrent exposure to higher levels of ozone results in chronic inflammation in the smallest airways, with some changes in the structure of these airways and the presence of increased collagen in the lung.[36-38] In my opinion, these results and the recent observation that inflammation occurs following exposure of humans to 0.40 ppm ozone[8] are causes for concern. The likelihood that individuals are being exposed to low level peaks of longer duration than one to two hours, and the possibility that these exposures may result in acute effects including inflammation of greater magnitude than those seen in two hour laboratory exposures at similar concentrations, are cause for even greater concern. Acute decrements in FEV_1 of 10% and more are unlikely to be due to measurement error, and represent the reaction of ozone in the airways. In my opinion, occasional decrements of this magnitude, or even somewhat larger, and the inflammation that may accompany them, are completely reversible if these exposures occur in isolation. Frequent exposures resulting in decrements of this magnitude or less frequent ones superimposed on longer-lasting lower levels of ozone, however, could potentially result in permanent lung damage.

Some have suggested that the phenomenon of attenuation of acute response with repeated exposures indicates that chronic effects of recurrent exposure are unlikely. The presence of continued inflammation and progressive changes in lung structure in animals makes it unlikely that attenuation of these neurally mediated effects reflects a protective effect for the potentially more serious effects of recurrent ozone exposure. In any case, it is ill-advised to assume this in the absence of supporting evidence.

REFERENCES

1. McDonnell, W. F., D. H. Horstmann, M. J. Hazucha, E. Seal, Jr., E. D. Haak, S. Salaam, and D. E. House. "Pulmonary Effects of Ozone Exposure During

Exercise: Dose-Response Characteristics." *J. Appl. Physiol.: Respir. Environ. Exercise Physiol.* 54:1345–1352 (1983).

2. Kulle, T. J., L. R. Sauder, J. R. Hebel, and M. D. Chatham. "Ozone Response Relationships in Healthy Nonsmokers." *Am. Rev. Respir. Dis.* 132:36–41 (1985).

3. Gibbons, S. I., and W. C. Adams. "Combined Effects of Ozone Exposure and Ambient Heat on Exercising Females." *J. Appl. Physiol: Respir. Environ. Exercise Physiol.* 57:450–456 (1984).

4. Folinsbee, L. J., B. L. Drinkwater, J. F. Bedi, and S. M. Horvath. "The Influence of Exercise on the Pulmonary Changes Due to Exposure to Low Concentrations of Ozone," in L. J. Folinsbee, J. A. Wagner, J. F. Borgia, B. L. Drinkwater, J. A. Gliner, and J. F. Bedi, eds. *Environmental Stress: Individual Human Adaptations.* (New York, NY: Academic Press, 1978), pp. 125–145.

5. Bates, D. V., G. M. Bell, C. D. Burnham, M. Hazucha, J. Mantha, L. D. Pengelly, and F. Silverman. "Short-Term Effects of Ozone on the Lung." *J. Appl. Physiol.* 32:176–181 (1972).

6. Beckett, W. S., W. F. McDonnell, D. H. Horstman, and D. E. House. "Role of the Parasympathetic Nervous System in the Acute Lung Response to Ozone." *J. Appl. Physiol.* 59:1879–1885 (1985).

7. Hazucha, M. J., D. V. Bates, and P. A. Bromberg. "Mechanism of Action of Ozone on the Human Lung." *Am. Rev. Respir. Dis.* 133:A214 (1986).

8. Seltzer, J., B. G. Bigby, M. Stulbarg, M. J. Holtzman, J. A. Nadel, I. F. Ueki, G. D. Leikauf, E. J. Goetzl, and H. A. Boushez. "O_3-Induced Change in Bronchial Reactivity to Methacholine and Airway Inflammation in Humans." *J. Appl. Physiol.* 60:1321–1326 (1986).

9. Holtzman, M. I., J. H. Cunningham, J. R. Sheller, G. B. Irsigler, J. A. Nadel, and H. A. Boushey. "Effect of Ozone on Bronchial Reactivity in Atopic and Nonatopic Subjects." *Am. Rev. Respir. Dis.* 120:1059–1067 (1979).

10. Kehrl, H. R., L. M. Vincent, R. J. Kowalsky, D. H. Horstman, J. J. O'Neil, W. H. McCartney, and P.A. Bromberg. "Ozone Exposure Increases Respiratory Epithelial Permeability in Humans." *Am. Rev. Respir. Dis.* 135:1124–1128 (1987).

11. Silverman, F., L. J. Folinsbee, J. Barnard, and R. J. Shephard. "Pulmonary Function Changes in Ozone—Interaction of Concentration and Ventilation." *J. Appl. Physiol.* 41:859–864 (1976).

12. Adams, W. C., W. M. Savin, and H. E. Christo. "Detection of Ozone Toxicity During Continuous Exercise Via the Effective Dose Concept." *J. Appl. Physiol.* 51:415–422 (1981).

13. Avol, E. L., W. S. Linn, T. G. Venet, D. A. Shamoo, and J. D. Hackney. "Comparative Respiratory Effects of Ozone and Ambient Oxidant Pollution Exposure During Heavy Exercise." *J. Air Pollut. Control Assoc.* 34:804–809 (1984).

14. McDonnell, W. F., III, D. H. Horstman, S. Abdul-Salaam, and D. E. House. "Reproducibility of Individual Responses to Ozone Exposure." *Am. Rev. Respir. Dis.* 131:36–40 (1985).

15. Gliner, J. A., S. M. Horvath, and L. J. Folinsbee. "Pre-exposure to Low Ozone Concentrations Does Not Diminish the Pulmonary Function Response on Exposure to Higher Ozone Concentration." *Am. Rev. Respir. Dis.* 127:51–55 (1983).

16. Gerrity, T. R., R. A. Weaver, J. H. Berntsen, and J. J. O'Neil. "Nasopharyngeal and Lung Removal of Ozone During Tidal Breathing in Man." *The Physiologist* 29:173 (1986).

17. McDonnell, W. F., D. H. Horstman, S. Abdul-Salaam, L. J. Raggio, and J. A. Green. "The Respiratory Responses of Subjects with Allergic Rhinitis to Ozone Exposure and Their Relationship to Nonspecific Airway Reactivity." *Toxicology and Industrial Health* 3:507–517 (1987).

18. McDonnell, W. F., III, R. S. Chapman, M. W. Leigh, G. L. Strope, and A. M. Collier. "Respiratory Responses of Vigorously Exercising Children to 0.12 ppm Ozone Exposure." *Am. Rev. Respir. Dis.* 132:876–879 (1985).

19. Avol, E. L., W. S. Linn, D. A. Shamoo, L. M. Valencia, U. T. Anzar, and J. D. Hackney. "Respiratory Effects of Photochemical Oxidant Air Pollution in Exercising Adolescents." *Am. Rev. Respir. Dis.* 132:875–879 (1985).

20. DeLucia, A. J., J. A. Whitaker, and L. R. Bryant. "Effects of Combined Exposure to Ozone and Carbon Monoxide in Humans," in M. A. Mehlman, S. D. Lee, and M. G. Mustafa, eds. *International Symposium on the Biomedical Effects of Ozone and Related Photochemical Oxidants*, March 1982, Pinehurst, NC. (Princeton, NJ: Princeton Scientific Publishers, Inc., 1983), pp. 145–159.

21. Lauritzen, S. K., and W. C. Adams. "Ozone Inhalation Effects Consequent to Continuous Exercise in Females: Comparison to Males." *J. Appl. Physiol.* 59:1601–1606 (1985).

22. Koenig, J. Q., D. S. Covert, M. S. Morgan, M. Horike, N. Horike, S. G. Marshall, and W. E. Pierson. "Acute Effects of 0.12 ppm Ozone or 0.12 ppm Nitrogen Dioxide on Pulmonary Function in Healthy and Asthmatic Adolescents." *Am. Rev. Respir. Dis.* 132:648–651 (1985).

23. Linn, W. S., R. D. Buckley, C. E. Spier, R. L. Blessey, M. P. Jones, D. A. Fischer, and J. D. Hackney. "Health Effects of Ozone Exposure in Asthmatics." *Am. Rev. Respir. Dis.* 117:835–843 (1978).

24. Silverman, F. "Asthma and Respiratory Irritants (Ozone)." *Environ. Health Perspect.* 29:131–136 (1979).

25. Linn, W. S., D. A. Shamoo, T. G. Venet, C. E. Spier, L. M. Valencia, U. T. Anzar, and J. D. Hackney "Response to Ozone in Volunteers with Chronic Obstructive Pulmonary Disease." *Arch. Environ. Health* 38:278–283 (1983).

26. Solic, J. J., M. J. Hazucha, and P. A. Bromberg. "The Acute Effects of 0.2 ppm Ozone in Patients with Chronic Obstructive Pulmonary Disease." *Am. Rev. Respir. Dis.* 125:664–669 (1982).

27. Kehrl, H. R., M. J. Hazucha, J. J. Solic, and P. A. Bromberg. "Responses of Subjects with Chronic Pulmonary Disease After Exposures to 0.3 ppm Ozone."*Am. Rev. Respir. Dis.* 131:719–724 (1985).

28. Hackney, J. D., W. S. Linn, J. G. Mohler, and C. R. Collier. "Adaptation to Short-Term Respiratory Effects of Ozone in Men Exposed Repeatedly." *J. Appl. Physiol.: Respir. Environ. Exercise Physiol.* 43:82–85 (1977).

29. Folinsbee, L. J., J. F. Bedi, and S. M. Horvath. "Respiratory Responses in Humans Repeatedly Exposed to Low Concentrations of Ozone." *Am. Rev. Respir. Dis.* 121:431–439 (1980).

30. Horvath, S. M., J. A. Gliner, and L. J. Folinsbee. "Adaptation to Ozone: Duration of Effect." *Am. Rev. Respir. Dis.* 123:496–499 (1981).

31. Kulle, T. J., L. R. Sauder, H. D. Kerr, B. P. Farrell, M. S. Bermel, and D. M. Smith,. "Duration of Pulmonary Function Adaptation to Ozone in Humans." *Am. Ind. Hyg. Assoc. J.* 43:832–837 (1982).

32. Spektor, D. M., M. Lippman, G. Thurston, F. E. Spizer, and C. Hayes. "Effects of Ambient Ozone on Respiratory Function in Active Healthy Children." *Am. Rev. Respir. Dis.* 135:A56 (1987).
33. Folinsbee, L. J., D. H. Horstman, P. J. Ives, S. A. Salaam, and W. F. McDonnell. "Respiratory Effects of Prolonged Exposure to 0.12 ppm Ozone." *Am. Rev. Respir. Dis.* 135:A57 (1987).
34. Whittemore, A. S., and E. L. Korn. "Asthma and Air Pollution in the Los Angeles Area." *Am. J. Public Health* 70:687–696 (1980).
35. Holguin, A. H., P. A. Buffler, C. F. Contant, Jr., T. H. Stock, D. Kotchmar, B. P. Hsi, D. E. Jenkins, B. M. Gehan, L. M. Noel, and M. Mei. "The Effects of Ozone on Asthmatics in the Houston Area," in S. D. Lee, ed. *Evaluation of the Scientific Basis for Ozone/Oxidants Standards*, November 1984, Houston, TX. (Pittsburgh, PA: Air Pollution Control Association, 1985), pp. 262–280.
36. Gross, K. B., and H. J. White. "Pulmonary Functional and Morphological Changes Induced by a 4-Week Exposure to 0.7 ppm Ozone Followed by a 9-Week Recovery Period," *Journal of Toxicology and Environmental Health.* 17:143–157 (1986).
37. Moore, P. F., and L. W. Schwartz. "Morphological Effects of Prolonged Exposure to Ozone and Sulfuric Acid Aerosol on the Rat Lung," *Exp. Mol. Pathol.* 35:108–123 (1981).
38. Eustis, S. L., L. W. Schwartz, P. C. Kosch, and D. L. Dungworth. "Chronic Bronchiolitis in Nonhuman Primates After Prolonged Ozone Exposure." *Am. J. Pathol.* 105:121–137 (1981).

CHAPTER 9

Extrapulmonary Effects of Ozone

Andrew T. Canada and Edward J. Calabrese

INTRODUCTION

Although the majority of studies which have reported on the adverse effects of ozone have studied the primary target organ, the lung, ozone has been reported to produce a number of effects on organ systems other than the lung. It is unlikely that these effects can be explained by the absorption of ozone into the blood via inhalation. Ozone is a strong oxidant and there is no evidence that it can cross the pulmonary epithelium and be absorbed into the systemic circulation. The number of studies reporting extrapulmonary effects not only are extensive but also involve multiple organ systems, many distant from the lung. Although the mechanism by which the effects on organ systems other than the lung is not known, it is likely that such effects are mediated by products of oxidant stress originating in the lung.

Hematologic Effects

Ozone has been reported to produce a number of varied effects on the hematologic system, both on the formed elements and on serum (plasma) biochemistry.

A. Erythrocyte

One of the earliest reported effects of ozone on the erythrocytes (RBCs) was the report of Brinkman and Lamberts,[1] who exposed volunteers to 1 ppm × 10 min and measured the disassociation of oxygen from hemoglobin following occlusion of blood supply to a finger. Their data showed that following ozone, the disassociation of oxygen from hemoglobin was incomplete (40%) compared with controls which showed complete disassociation.

Although Chow et al.[2] did not observe any changes in the biochemistry of RBCs taken from rats and monkeys, Buckley et al.[3] reported a number of changes in the morphology and biochemistry of RBCs taken from human volunteers exposed to 0.5 ppm × 2.75 hr. Among these were a 20% increase in fragility, a 20% decrease in acetylcholine esterase activity, a 15% decrease in reduced glutathione (GSH), and a small but statistically significant increase in intracellular lactic acid dehydrogenase (LDH) activity. Hackney et al.[4] compared the biochemical and physiological effects of exposure to 0.37 ppm × 2 hr on a group of Canadians from an area of low air pollution with a group of volunteers from Los Angeles, presumably exposed to high levels of air pollutants, including ozone. Among the variables evaluated were erythrocyte fragility and acetylcholinesterase activity. The volunteers with no previous exposure to air pollution had a 34% to 53% increase in RBC fragility and a 7% to 15% decrease in erythrocyte acetylcholinesterase activity. The reduction in acetylcholine esterase has also been reported by Jaffe[5] who found a 50% decrease in the erythrocytes of mice exposed to 8 ppm × 8 hr, a very high exposure. The survival of erythrocytes in rabbits exposed to 0.06 to 0.48 ppm for 2.75 hr was measured by the release of ^{51}Cr measured 24 hr following the exposure.[6] A decreased survival of the labeled erythrocytes was reported for all exposures excepting the 0.20 ppm concentration. A possible reason why the observed difference at 0.24 ppm did not reach statistical significance may have been due to the heterogeneity of commercially bred rabbits.

Goldstein[7] investigated the proposal that inhaled ozone would result in increased production of oxygen free radicals of which hydrogen peroxide would be a detectable oxidant metabolite. Following an in vivo exposure of rats to 5 ppm and mice to 6.7 ppm, RBC catalase activity decreased by 50%, suggesting the production of intracellular hydrogen peroxide.

Exploring the hypothesis that ozone could affect the RBC membrane, rats were exposed to 0.5 to 2.0 ppm for 1 hr.[8] RBCs taken from the exposed rats showed decreased agglutination when incubated with concanavalin A, a reagent used as a probe for plasma membrane changes. The loss of membrane stability has also been shown by Lamberts and Veninga,[9] who found increased sphering of RBCs taken from mice, rabbits, and humans exposed to 0.25 ppm for 1 to 6 hr. A decrease in the deformability of RBCs from mice also has been observed after ozone exposure, 0.3 to 1.0 ppm × 4 hr.[10] These collective findings may be significant since RBCs must be able to alter their shape to pass through capillaries normally too small to accommodate their normal diameter.

Using the Dorset sheep, a species low in G-6-phosphate dehydrogenase (G-6-PD) as a model for human G-6-PD deficiency, Moore et al.[11] found that 0.7 ppm × 2.75 hr produced an increase in methemoglobin and a decrease in the number of red blood cells. In a later study, this group also found decreased RBC survival with ozone in the same model.[12] Both guinea pigs and rats who survived a very high exposure to ozone had an increase in hemoglobin and hematocrit, likely a compensatory response to the acute reduction of total RBCs reported previously.[13] The ability of vitamin E to protect against a number of the pulmonary effects of ozone is well known.[14,15] In order to evaluate the effect of vitamin E in protecting against ozone-induced damage to the RBC, Chow and Kaneko[16] fed rats a vitamin E deficient diet and then exposed them to 0.8 ppm of ozone continuously for seven days. Under those conditions, they found an increase in RBC glutathione peroxidase activity, pyruvate kinase, and LDH while the glutathione content decreased. Lastly, Arkin et al.[17] investigated the biochemistry of RBCs isolated from rats exposed to 1.5 ppm for three days at 6 and 8 ppm for 4 hr. No effect was observed on potassium influx, glutathione peroxidase, or superoxide dismutase. However, there was a change in shape of the RBCs with a number of transformations from the normal discocytes to echinocytes observed.

B. Serum

Human volunteers exposed to 0.5 ppm × 2.5 hr had a 50% increase in the concentration of malondialdehyde (MDA) like-reacting substances apparently reflecting the oxidant stress on the lung and subsequent appearance of the MDA-like substances into the serum.[3] Those investigators also reported an increase in serum vitamin E and a decrease in glutathione reductase concentrations. A similar study[4] found a small but statistically significant increase in serum vitamin E concentration in volunteers exposed to 0.37 ppm for 2 hr with four 15 min exercise periods. An animal study investigating the effects of 3.5 ppm × 4 hr over ten days, reported a reduction in the serum concentration of vitamin E in the ozone group, while no change was observed in air controls.[18] In a rabbit study investigating the effect of ozone, 0.3 ppm for 3.75 hr on the plasma concentrations of vitamins A, C, and E, Canada et al.[19] failed to detect any changes in any of the vitamins. Thus, whether a moderate exposure to ozone can produce significant changes in the plasma (or serum) concentrations of vitamins participating as antioxidants remains an open question.

Following exposure of rabbits to a high dose of ozone, 10 ppm × 1 hr, an increase in serum alkaline phosphatase was reported.[20] The tissue source of this enzyme was not investigated. Presumably, it reflected massive damage to pulmonary tissue with subsequent leakage of enzyme into the blood. Reflecting the severity of the damage, the serum alkaline phosphatase activity did not return to normal for one week following the exposure.

Another group exposed rats to 4 ppm for 2, 4, and 8 hr and reported an increase in the plasma concentrations of the prostaglandins, PGF2a and PGE2.[21] The observed increase for PGF2a and PGE2 was the greatest following the 2 hr exposure, 186% and 220%, respectively. The increase in prostaglandins likely represents an oxidant-induced mobilization of arachidonate from the plasma membrane.

Veninga et al.[22] found that the serum creatine kinase [CPK] increased in mice exposed to 0.02 ppm of ozone for two hours. In a later study,[23] this group reported similar results for rats exposed to a regimen similar to the mice. Strangely, with exposure to higher concentrations, there was no increase in CPK. The authors' explanation that this represented an adaptive response is less than satisfactory as adaptation usually requires a previous exposure or continuous nonlethal exposure.

One of the earliest reports of an effect of ozone on serum lipids was that of Shimasaki et al. in 1976.[24] They found that after exposing rats to 1.1 ppm continuously for 24 hours, both the lecithin:cholesterol acyl transferase activity and free cholesterol increased significantly. An effect of inhaled ozone on serum lipids was also detected in a study where rats were exposed to 1, 1.75, and 3 ppm $\times$ 5 hr for ten days.[25] An ozone-related increase was noted in the lipid components, free cholesterol and high density lipoproteins. However, serum triglycerides tended to decrease as a result of the ozone exposure. Whether the same effect would be seen in humans exposed to ozone and if so, whether the observed changes have clinical significance remains to be shown. In a similar study, except with monkeys as the experimental animal, Rao et al.[26] found that in addition to a decrease in lung polyunsaturated fatty acids, lecithin cholesterol acyl transferase was increased in monkeys exposed to 0.3 ppm 8 hr daily for 90 days. A sex-related difference in the effect of ozone on serum cholesterol and lipoproteins was reported by Vaughan et al.[27] The increase in both of these serum constituents occurred only in male guinea pigs exposed to 1 ppm continuously for 14 days.

Neurologic Effects

Another organ system affected by ozone is the central nervous system. In humans, exposure to concentrations greater than 0.6 ppm have been reported to result in lethargy and severe headaches.[9,28,29] In the case of severe poisonings, a headache was one of the earliest symptoms, preceded only by a dry cough.[20] A large epidemiologic study found a relationship between the levels of oxidant pollution and the occurrence of headaches in a group of student nurses.[30]

Ozone has also been shown to have an effect on a number of visual measurements. Volunteers exposed to concentrations as low as 0.2 ppm showed a decrease of visual acuity in the scotopic and mesopic ranges, an increase in peripheral vision, and changes in the balance of the extraocular muscles.[31]

A number of behavioral studies in rodents have demonstrated a dose-response relationship for ozone. Weiss et al.[32] reported that rats exposed to ozone concentrations ranging from 0.1 to 2.0 ppm had a decreased response to a fixed interval feeding regimen, a sensitive measurement of toxicant exposure. Extending their studies, this group then found that ozone produced a decrease in the amount of running time of a rat allowed free access to a running wheel.[33] This decrease could be attributed to longer rest periods between exercising than observed with the same rats during a pre-ozone control period. This proved to be a highly sensitive indicator of an ozone effect, since the increased respirations resulting from running increased the total exposure to ozone, producing an effect at a concentration as low as 0.2 ppm.

A decrease in the cerebral concentrations of catechol-o-methyltransferase was observed in dogs exposed to 8, 12, or 16 hr of ozone at a concentration of 1 ppm.[34] This reduction in enzyme activity correlated with a decreased catecholamine content of the brain. The effect of this decrease in catecholamine concentration on cerebral function is unknown.

Hepatic Effects

Although no data were presented, Murphy et al.[20] reported that the liver content of alkaline phosphatase in rats exposed to 3.1 ppm for 20 hr increased by 50%. In another report where rabbits were exposed to 10 ppm × 1 hr for one day, an initial decrease in DNA and RNA occurred during the first four hours after exposure, with both returning to normal by 24 hr.[35] No potential explanation for these findings was offered.

The first report of an ozone effect on hepatic levels of ascorbic acid was that of Veninga et al.[22] They showed that in mice exposed to ozone, liver concentrations of ascorbic acid increased by 10% to 20%. Dubick et al.[36] also investigated the effect of ozone on tissue ascorbic acid concentration in mice continuously exposed to 1.5 ppm ozone for 5 days. They found the hepatic concentration decreased transiently. However, this decrease was only statistically significant during the first 24 hr of the exposure. They attributed this decrease to a redistribution of ascorbic acid from the liver to the lung, where the ascorbic acid had been depleted by the oxidant stress. In support of this explanation, they did show a concomitant decrease in lung ascorbic acid. Continuing their studies using [^{14}C] ascorbic acid, they were not able to show an effect of ozone on the decay curve of ascorbic acid in lung, liver, carcass, or serum. There is no apparent explanation for the discrepancy in the hepatic changes observed in ascorbic acid content between the two studies.

In 1974, Gardner et al.[37] exposed female mice to ozone, 1 ppm × 3 hr for seven consecutive days. They then measured the effect of this on pentobarbital sleeping time, defined as the time following an injection of pentobarbital from when the animals could be placed on their backs without attempting to arise, to when they spontaneously righted. They found that by Day 2, this increased from 20.2 min to 35.3 min. A statistically insignificant increase at this point also was seen in controls. By exposure Day 4, there was no longer any difference between the ozone-exposed and the air control mice. As the pentobarbital sleeping time is used to reflect general drug oxidative metabolizing activity, these results were interpreted as suggesting an effect of ozone on hepatic drug metabolism.

Continuing this work, Graham et al.[38] were able to show the ozone-induced increase in pentobarbital sleeping time was also seen in female mice (three strains), rats, and hamsters. No or a very small effect was observed for males of each of the species. No explanation was proposed for the observed sex-related difference in ozone response. They found a relationship between exposure time (T) in hr and concentration (C) in ppm such as that when $T \times C = 4.5$, the prolongation of sleeping time was observed. They suggested that a product of lipid peroxidation resulting from oxidant stress on the lung somehow might impair hepatic P-450 metabolizing enzymes. Based on this hypothesis, they then investigated the activity of some of the hepatic microsomal enzymes harvested from female mice exposed to ozone.[39] Surprisingly, no effect was observed on either aminopyrine N-demethylase or p-nitroanisole O-demethylase. They also reported an increase in aniline hydroxylase activity. These data do not support the explanation that the decrease in pentobarbital sleeping time was due to an ozone effect on the liver, although a later study[40] by the same group clearly showed the increase in sleeping time was entirely due to an increase in the pentobarbital serum half-life. Providing support for the work of Gardner and Graham, Takahashi et al.,[41] exposed rats to a higher dose than that of either Gardner et al.[37] or Graham et al.[38] (0.8 ppm for 7 hr), and reported a significant decrease in the activity of a number of liver metabolizing enzymes. Among these were benzopyrene hydroxylase, 7-ethoxycoumarin 0-deethylase, and aniline hydroxylase. No effect was noted on p-nitroanisole N-demethylase activity. A decrease of approximately 20% also was observed in microsomal protein, cytochrome P-450, NADPH cytochrome P-450 reductase, cytochrome b5, and NADH-cytochrome b5 reductase. In an experiment designed to investigate the effect of animal age on the ozone-inhibition of drug metabolism, Canada and Calabrese[42] did not find an inhibition of theophylline metabolism in young rabbits. However, they did observe that after exposure to 0.3 ppm for 3.75 hr, mature rabbits (2–4 yr of age) did exhibit an approximate 50% increase in the serum half-life of theophylline. This led them to suggest that the prolongation of theophylline half-life in the mature, but not in the young rabbits, represented an age-related effect of oxidant stress. In a later study following the same protocol as Graham et al.[38], these same investigators did not find a prolongation of sleeping time in young (3 mo), but did observe a prolongation in mature (18 mo), female mice.[43] They attributed the increase in pentobarbital sleeping time as well as the increase in theophylline half-life to an effect on the P-450 metabolizing system in the lung, rather than in the liver. In support of this hypothesis that the effect of ozone on P-450 metabolism was restricted to the lung was the finding that following exposure of rabbits to 1.0 ppm for 1.5 hr, a 50% decrease in lung P-450 content was observed.[44]

Ozone, 1.5 ppm × 30 to 60 min, produced an increase in fluorescent pigments in the liver.[45] These pigments have been postulated to be end products of lipid peroxidation. It is interesting that despite an increase in these products in the liver, no similar increase was found in the primary target tissue, the lung.

Mutagenic

In the first study to evaluate the potential for inhaled ozone to be a mutagen, Zelac et al.[46] exposed hamsters to 0.2 ppm $\times$ 5 hr and peripheral lymphocytes were evaluated for chromosomal aberrations. With exposure-adjusted break frequencies as the measured variable, an ozone-induced increase was observed. In another study,[47] this group showed ozone to be a more potent mutant than radiation. Following the reports of Zelac, Merz et al.[48] decided to investigate the effect of 0.5 ppm ozone exposure for either 8 or 10 hr on chromatid and chromosomal aberrations in six normal human volunteers. They did not find any chromosome-type aberrations, however, but did report an increase in both achromatic and chromatid deletions which persisted as long as two weeks post-exposure. These results, presented as preliminary data, have not been published in their entirety. Using a lower dose (0.4 ppm for 4 hr) in 30 male human volunteers, McKenzie et al.[49] did not detect any ozone-induced increase in chromosomal aberrations or in chromatid-chromosome breaks of lymphocytes.

Cardiovascular

Ozone has also been reported to produce a number of effects on the cardiovascular system. One to four minutes following exposure to a high concentration of ozone (50 ppm) tracheotomized dogs underwent a period of apnea accompanied by hypotension and bradycardia.[50] These effects were mitigated by vagotomy showing that they were likely due to stimulation of the parasympathetic vagus nerve as a result of the irritant effect of the high ozone concentration. Lamberts and Veninga[9] exposed rabbits to 0.2 ppm for 5 hours $\times$ 4 weeks and looked for pathological changes within the myocardium. They found ruptures of the nuclear envelope with extrusion of the nuclear contents into the cytoplasm of the myocyte.

Endocrine Effects

Ozone has also been shown to produce multiple effects on the endocrine system. The most prominent among these is an effect on the pituitary/thyroid axis. Fairchild et al.[51] reported that single 5-hour exposures to 1, 2, and 4 ppm of ozone resulted in a pronounced inhibition of the release of ^{131}I, an effect first detected two days after the exposure and which continued to be detected at day 12. Carrying these investigations further, Clemons and Garcia[52] reported that exposure of rats to 1 ppm for 24 hr reduced plasma thyrotropin by 50%, reduced the levels of both T3 and T4 without having any effect on T3 uptake, and decreased the PBI from 4.9 to 2.9 μg/100mL. Simultaneously, the concentration of prolactin increased. The reduction of thyrotropin activity was found to be sensitive to time $\times$ concentration effect with titration of either component (T or C) yielding similar effects on the pituitary thyroid axis. These investigators also reported an ozone-induced increase in the weight of the thyroid gland. Since the administration of thyroid hormone had been reported to potentiate and thyroid blocking agents to protect against the lethality of ozone,[53] the authors felt the decrease in thyroid hormone activity represented a protective mechanism to reduce the ozone-induced damage to the animal.

Another surprising effect of ozone on the parathyroid gland has been reported by Atwal and Wilson.[54] Rabbits were exposed to 0.75 ppm for four or eight hours and the parathyroid gland investigated histopathologically. A number of significant changes were observed in the ozone treated animals which were not present in controls. Among these were: production of a large number of secretion granules, hyperplasia of the chief cells, proliferation and hyperplasia of the rough endoplasmic reticulum, ribosomes, mitochondria, Golgi complex, lipid bodies, and accumulation of secretion granules inside the vascular endothelium. In a later study,[55] the parathyroid glands of dogs exposed to 0.75 ppm for 48 hr showed the appearance of atypical cilia in ciliated cysts of the parathyroid gland. As no study has investigated parathyroid hormonal changes which may occur with ozone treatment, the significance of the observed parathyroid changes to humans exposed to ozone remains to be determined.

Other Effects

The effect of inhaled ozone on cholinesterase activity has also been reported for tissues other than blood. Gordon et al.[56] reported that 0.8 ppm $\times$ 1 hr produced a 14% decrease in the cholinesterase activity of the diaphragm of guinea pigs compared with a 16% decrease in the lung.

A potential effect on the immune response system was suggested by the work of Eskew et al.[57] who found that spleen cells isolated from selenium-deficient rats exposed continuously to 1 ppm $\times$ 8 hr for 7 days had an enhanced ability to mediate antibody-dependent cell-mediated cytotoxicity. Another study investigated the effect of ozone on cell mediated immunity in humans exposed to 0.6 ppm for 2 hr.[58] Although there was no effect on the number of T lymphocytes observed, the blastogenic response to phytohemagglutin was depressed two and four weeks after exposure.

Ozone has also been shown to be potentially fetotoxic. Following exposure of female mice to 0.2 ppm $\times$ 7 hr for 3 weeks, there was an increase in fetal mortality from 7.5% for controls to 34% for the exposed group.[9] Nonspecific inflammatory changes such as focal necrosis and periportal lymphocytic infiltrates have been observed in the livers from rats, guinea pigs, dogs, and monkeys exposed to ozone.[13] In the same study, kidneys showed cystic dilation of tubules and scattered pigmented tubular casts.

Female mice were exposed to 0.31 ppm continuously for 103 hr every other week for six months.[59] The exposed animals had a greater spleen weight, and spleen to body weight ratio greater than air-breathing controls. As with the majority of extrapulmonary effects, the authors offered no explanation for their findings.

CONCLUSION

Ozone has been reported to produce a variety of effects on organs other than the lung, the obvious target tissue. These effects have ranged from alterations in blood chemistry which could be explained by leakage of damaged pulmonary cell contents into the blood stream, to other effects such as those on cholesterol and the parathyroid gland which defy logical explanation. Finally, the importance of any of the observed extrapulmonary effects to the toxicology of ozone to humans remains to be shown.

REFERENCES

1. Brinkman, R., and H. B. Lamberts. "Ozone as a Possible Radiometric Gas," *Nature* 181:1202–1203 (1958).
2. Chow. C., M. Mustafa, C. Cross, and B. Tarkington. "Effect of Ozone Exposure on the Lungs and the Erythrocytes of Rats and Monkeys: Relative Biochemical Changes," *Environ. Physiol. Biochem.* 5:142–148 (1975).
3. Buckley, R., J. Hackney, K. Clark, and C. Posin. "Ozone and Human Blood," *Arch. Environ. Health* 30:40–43 (1975).
4. Hackney, J., W. Linn, S. Karuza, R. Bulkley, D. Law, D. Bates, M. Hazucha, L. Pengelly, and F. Silverman. "Effects of Ozone Exposure in Canadians and Southern Californians," *Arch. Environ. Health* 32:110–116 (1977).
5. Jaffe. L. S. "Photochemical Air Pollutants and Their Effects on Man and Animals," *Arch. Environ. Health* 16:241–255.
6. Calabrese, E., G. Moore, and E. Grunwald. "Ozone-Induced Decrease of Erythrocyte Survival in Adult Rabbits," in *The Biomedical Effects of Ozone and Related Photochemical Oxidants* (Princeton: Princeton Scientific Publishers, 1983), p. 103.
7. Goldstein, B. D. "Hydrogen Peroxide in Erythrocytes," *Arch. Environ. Health* 26:279–280 (1973).
8. Hamberger, S. J., and B. D. Goldstein. "Effect of Ozone on the Agglutination of Erythrocytes by Concanavalin A," *Environ. Res.* 19:292–298 (1979).
9. Lamberts, H. B., and T. S. Veninga. "Radiomimetic Toxicity of Ozonized Air," *Lancet* 1:133–136 (1964).
10. Morgan, D., Dorsey, A., and D. Menzel. "Erythrocytes from Ozone-Exposed Mice Exhibit Decreased Deformability," *Fundam. Appl. Toxicol.* 5:137–143 (1985).
11. Moore, G., E. Calabrese, and E. Schulz. "Effect of In Vivo Ozone Exposure to Dorset Sheep, an Animal Model with Low Levels of Erythrocyte Glucose-6-Phosphate Dehydrogenase Activity," *Bull. Environ. Contam. Toxicol.* 26:273–280 (1981).
12. Moore, G., E. Calabrese, and F. Labato. "Erythrocyte Survival in Sheep Exposed to Ozone," *Bull. Environ. Contam. Toxicol.* 27:26–138 (1981).
13. Jones, R., L. Jenkins, R. Coon, and J. Siegel. "Effects of Long-Term Continuous Inhalation of Ozone on Experimental Animals," *Toxicol. Appl. Pharmacol.* 17:189–202 (1970).
14. Chow, C. K., and A. L. Tappel. "Activities of Pentose Shunt and Glycolytic Enzymes in Lungs of Ozone-Exposed Rats, *Arch. Environ. Health* 26:205–208 (1973).
15. Sato, S., M. Kawakami, S. Maeda, and T. Takishima. "Scanning Electron Microscopy of the Lungs of Vitamin E-Deficient Rats Exposed to a Low Concentration of Ozone," *Am. Rev. Resp. Dis.* 113:809–821 (1976).

16. Chow, C. K., and J. J. Kaneko. "Influence of Dietary Vitamin E on the Red Cells of Ozone-Exposed Rats," *Environ. Res.* 19:49–55 (1979).
17. Larkin, E., S. Kimzey, and K. Siler. "Response of the Rat Erythrocyte to Ozone Exposure," *J. Appl. Physiol.* 45(6):893–898 (1978).
18. Goldstein, B., R. Buckley, R. Cardenas, and O. Balchum. "Ozone and Vitamin E," *Science* 169:605–606 (1970).
19. Canada, A., C. Chow, G. Airriess, and E. Calabrese. "Lack of Ozone Effect on Plasma Concentrations of Retinol, Ascorbic Acid, and Tocopherol," *Nutrit. Res.* 7:797–799 (1987).
20. Murphy, S., H. Davis, and V. Zaratzian. "Biochemical Effects in Rats from Irritating Air Contaminants," *Toxicol. Appl. Pharmacol.* 6:520–528 (1964).
21. Giri, S., M. Hollinger, and M. Schiedt. "The Effects of Ozone and Paraquat on PGF_{2a} and PGE_2. Levels in Plasma and Combined Pleural Effusion and Lung Lavage of Rats," *Environ. Res.* 21:467–476 (1980).
22. Veninga, T., J. Wagenaar, and W. Lemstra. "Distinct Enzymatic Responses in Mice Exposed to a Range of Low Doses of Ozone," *Environ. Health Perspect.* 39:153–157 (1981).
23. Veninga, T. S., and V. Fidler. "Ozone-Induced Elevation of Creatine Kinase Activity in Blood Plasma of Rats," *Environ. Res.* 41:168–173 (1986).
24. Shimasaki, H., T. Takatori, W. Anderson, H. Horton, and O. Privett. "Alteration of Lung Lipids ln Ozone Exposed Rats," *Biochem. Biophys. Res. Comm.* 68:1256–1262 (1976).
25. Mole, M.. A. Stead, D. Gardner, F. Miller, and J. Graham. "Effect of Ozone on Serum Lipids and Lipoproteins in the Rat," *Toxicol. Appl. Pharmacol.* 80:367–376 (1985).
26. Rao, G., E. Larkin, J. Harkema, and D. Dungworth. "Changes in the Levels of Polyunsaturated Fatty Acids in the Lung and Lecithin Cholesterol Acyl Transferase Activity in Plasma of Monkeys Exposed to Ambient Levels of Ozone," *Toxicol. Lett.* 24:125–129 (1985).
27. Vaughan, W., G. Adamson, F. Lindgren, and J. Schooley. "Serum Lipid and Lipoprotein Concentrations Following Exposure to Ozone," *J. Environ. Pathol. Toxicol. Oncol.* 5:165–173 (1984).
28. Kleinfeld, M., and C. P. Giel. "Clinical Manifestations of Ozone Poisoning: Report of a New Source of Exposure," *Am. J. Med. Sci.* 231:638–643 (1956).
29. Kelly, F. J., and W. E. Gill. "Ozone Poisoning," *Arch. Environ. Health* 10:517–519 (1965).
30. Hammer, D., V. Hasselblad, B. Portnoy, and P. Wehrle. "Los Angeles Student Nurse Study, Daily Symptom Reporting and Photochemical Oxidants," *Arch. Environ. Health* 28:255–260 (1974).
31. Lagerwerff, J. M. "Prolonged Ozone Inhalation and Its Effects on Visual Parameters," *Aerospace Med.* 34:479–486 (1963).
32. Weiss, B., J. Ferin, W. Merigan, S. Stern, and C. Cox. "Modification of Rat Operant Behavior by Ozone Exposure," *Toxicol. Appl. Pharmacol.* 58:244–251 (1981).
33. Tepper, J., B. Weiss, and C. Cox. "Microanalysis of Ozone Depression of Motor Activity," *Toxicol. Appl. Pharmacol.* 64:317–326 (1982).
34. Trams, E., C. Lauter, E. Branderburger Brown, and O. Young. "Cerebral Cortical Metabolism After Chronic Exposure to Ozone," *Arch. Environ. Health* 24:153–159 (1972).

35. Scheel, L., O. Dobrogorski, J. Mountain, J. Svirbley, and H. Stokinger. "Physiologic, Biochemical, Immunological, and Pathologic Changes Following Ozone Exposure." *J. Appl. Physiol.* 14:67–80 (1959).

36. Dubick, M., J. Critchfield, J. Last, C. Cross, and R. Rucker. "Ascorbic Acid Turnover in the Mouse Following Acute Ozone Exposure," *Toxicology* 27:311–313 (1983).

37. Gardner, D., J. Illing, F. Miller, and D. Coffin. "The Effect of Ozone on Pentobarbital Sleeping Time in Mice," *Res. Comm. Chem. Pathol. Pharmacol.* 9(4):689–700 (1974).

38. Graham, J., D. Menzel, F. Miller, J. Illing, and D. Gardner. "Influence of Ozone on Pentobarbital-Induced Sleeping Time in Mice, Rats, and Hamsters," *Toxicol. Appl. Pharmacol.* 61:64–73 (1981).

39. Graham, J., F. Miller, D. Gardner, R. Ward, and D. Menzel. "Influence of Ozone and Nitrogen Dioxide on Hepatic Microsomal Enzymes in Mice," *J. Toxicol. Environ. Health* 9:849–856 (1982).

40. Graham, J., D. Menzel, M. Mole, F. Miller, and D. Gardner. "Influence of Ozone on Pentobarbital Pharmacokinetics in Mice," *Toxicol. Lett.* 24:163–170 (1985).

41. Takahashi, Y., T. Miura, and K. Kubota. "In Vivo Effect of Ozone Inhalation on Xenobiotic Metabolism of Lung and Liver of Rats," *J. Toxicol. Environ. Health* 15:855–864 (1985).

42. Canada, A. T., and E. J. Calabrese. "Age-Related Susceptibility of Ozone-Induced Inhibition of Theophylline Elimination," *Toxicol. Appl. Pharmacol.* 81:43–49 (1985).

43. Canada, A., E. Calabrese, and D. Leonard. "Age-Dependent Inhibition of Pentobarbital Sleeping Time by Ozone in Mice and Rats," *J. Gerontol.* 41:587–589 (1986).

44. Goldstein, B., S. Solomon, B. Pasternack, and D. Bickers. "Decrease in Rabbit Lung Microsomal Cytochrome P-450 Levels Following Ozone Exposure," *Res. Comm. Chem. Pathol. Pharmacol.* 10:759–762 (1975).

45. Csallany, A., J. Manwaring, and B. Menken. "Ozone-Related Fluorescent Compounds in Mouse Liver and Lung," *Environ. Res.* 37:320–326 (1985).

46. Zelac, R., H. Cromroy, W. Bolch, Jr., B. Dunavant, and H. Bevis. "Inhaled Ozone as a Mutagen. II. Effect on the Frequency of Chromosome Aberrations Observed in Irradiated Chinese Hamsters," *Environ. Res.* 4:325–342 (1971).

47. Zelac, R., H. Cromroy, W. Bloch, Jr., B. Dunavant, and H. Bevis. "Inhaled Ozone as a Mutagen. I. Chromosome Aberrations Induced in Chinese Hamster Lymphocytes." *Environ. Res.* 4:262–282 (1971).

48. Merz, T., M. Bender, H. Kerr, and T. Kulle. "Observations of Aberrations in Chromosomes of Lymphocytes from Human Subjects Exposed to Ozone at a Concentration of 0.5 ppm for 6 and 10 Hours," *Mutation Res.* 31:299–302 (1975).

49. McKenzie, W., J. Knelson, N. Rummo, and D. House. "Cytogenic Effects of Inhaled Ozone in Man, *Mutation Res.* 48:95–102 (1977).

50. Vaughan, T., W. Moorman, and T. Lewis. "Cardiopulmonary Effects of Acute Exposure to Ozone in the Dog," *Toxicol. Appl. Pharmacol.* 20:404–411 (1971).

51. Fairchild, E., S. Graham, M. Hite, R. Killens, and L. Scheel. "Changes in Thyroid I[131] Activity in Ozone-Tolerant and Ozone Susceptible Rats," *Toxicol. Appl. Pharmacol.* 6:607–613 (1964).

52. Clemons, G. K., and J. F. Garcia. "Endocrine Aspects of Ozone Exposure in Rats," *Arch. Toxicol.* Suppl. 4:301–304 (1980).

53. Fairchild, E. J., III, and S. L. Graham. "Thyroid Influence on the Toxicity of Respiratory Irritant Gases, Ozone and Nitrogen Dioxide," *J. Pharmacol. Exp. Ther.* 139:177–184 (1963).

54. Atwal, O. S., and T. Wilson. "Parathyroid Gland Changes Following Ozone Inhalation. A Morphologic Study," *Arch. Environ. Health* 28:91–100 (1974).
55. Pemsingh, R., O. Atwal, and R. MacPherson. "Atypical Cilia in Ciliated Cysts of the Parathyroid Glands of Dogs Exposed to Ozone," *Exp. Pathol.* 28:105–110 (1985).
56. Gordon, T., B. Taylor, and M. Amdur. "Ozone Inhibition of Tissue Cholinesterase in Guinea Pigs," *Arch. Envir. Health* 36(6):284–288 (1981).
57. Eskew, M., W. Scheuchenzuber, R. Scholz, C. Reddy, and A. Zarkower. "The Effects of Ozone Inhalation on the Immunological Response of Selenium- and Vitamin E-Deprived Rats. *Environ. Res.* 40:274–284 (1976).
58. Peterson, M., R. Smialowiz, S. Harder, B. Ketcham, and D. House. "The Effect of Controlled Ozone Exposure on Human Lymphocyte Function," *Environ. Res.* 24:299–308 (1981).
59. Hassett, C., M. Mustafa, W. Coulson, and R. Elashoff. "Splenomegaly in Mice Following Exposure to Ambient Levels of Ozone," *Toxicol. Lett.* 26:139–144 (1985).

CHAPTER 10

Epidemiologic Assessment of Short-Term Ozone Health Effects

Douglas W. Dockery and David Kriebel

INTRODUCTION

In this chapter we discuss the findings of recent epidemiologic studies of ozone. A recent EPA monograph presents an exhaustive review of this literature, and so no pretense of completeness will be offered.[1] We will focus on ozone's transient effects on pulmonary function and on the methodologic developments that have substantially improved our understanding of this problem.

Epidemiologic studies of the long-term health effects of photochemical oxidants at concentrations found in urban areas have been underway for many years. These studies have yet to establish convincing evidence for such chronic effects. For example, epidemiologic studies of the chronic effects of oxidant exposure in humans have reported significant differences in respiratory health between communities with differing exposures to photochemical oxidants.[2-4] These associations, while important, are difficult to interpret because of the difficulty of separating the effects of the various pollutants to which urban populations are inevitably exposed.

Epidemiologic studies of the acute effects of ozone, in contrast, have observed marked transient decrements in lung function and increased incidence of symptoms, but have suffered from a number of potential biases. Until recently these studies often did not account adequately for the correlation of individual responses from day to day, and the "drop-in" and "drop-out" of subjects who may be more or less sensitive to the effects of ozone. In addition, the measure of acute ozone exposure which appropriately characterizes physiologic response was not known, so that exposure may not be properly measured in order to evaluate health impact.

The quantification of these transient effects has benefited greatly from a number of chamber studies of volunteers. In these studies reductions in pulmonary function following controlled exposures to ozone alone have been observed consistently. For example, Kulle et al.[5] have shown that proportional declines in FEV_1 increase monotonically with increasing one-hour exposures to ozone concentrations between 100 and 250 ppb (Figure 1). Individual subjects appear to have

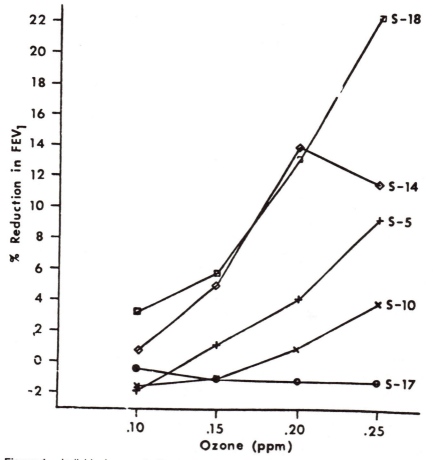

Figure 1.   Individual concentration-response plots (percent reduction from control) in FEV_1 with 2-hr chamber exposures to 0.10, 0.15, 0.20, and 0.25 ppm O_3. Four groups of observed responses were none (S-17 and S-10), mild to moderate (S-14), and moderate to severe (S-18).[5].

a characteristic response rate, which varies from small to steep slopes. These chamber studies have demonstrated the importance of considering respiratory ventilation rates in defining ozone dose delivered to the lungs. It appears that individuals within a sample population should be expected to have characteristic individual exposure response functions.

In epidemiologic studies, exposure is determined by factors outside of the experimenters control. To obtain complete data in which all subjects are exposed blindly to each exposure level is not possible. The analysis of epidemiologic studies is further complicated by missing observations. Korn and Whittemore[6] have shown how to design and analyze epidemiologic studies of a panel of subjects with repeated measurements of their individual response to air pollution exposure. Initially the exposure response function for each individual in the study is estimated. Population characteristics are then calculated from the individual exposure-response characteristics.

The perspective on physiologic response provided by the chamber studies, and the perspective on design and analysis provided by Korn and Whittemore, have provided guidance in the design of epidemiologic studies. In this decade, a series of epidemiologic studies have been published which have found a consistent decline of lung function associated with ozone exposures. Other studies have found associations between asthma attacks and ozone and also between hospital admissions for respiratory conditions and ozone. In this chapter, we will restrict our attention to these studies of transient effects which have been based on this perspective.

EPIDEMIOLOGIC STUDIES

Transient Effects on Pulmonary Function

The short-term, transient effects of ozone on pulmonary function have been reported by a number of investigators. A variety of measures of lung function have been reported, including the forced vital capacity (FVC), the forced expiratory volume in the first second (FEV_1), and flow rates at a range of different volumes: peak flow rate (PEFR), and maximum mid-expiratory flow (MMEF) being the most common. In this review we will concentrate on the FEV_1 and the PEFR, because ozone seems to affect these measures most strongly.

The general approach in these studies has been to make repeated measurements of pulmonary function on a sample of subjects, children or adults, along with concurrent measurements of ozone, and other pollution and meteorological variables. Each subject's pulmonary function is regressed against the ozone or other pollution measure for that time period, to determine a slope of response versus exposure, as in Figure 2. These slopes are then analyzed to determine if there is an effect overall or within specific sub-groups.

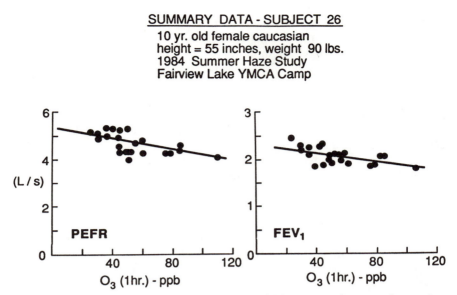

SUMMARY DATA - SUBJECT 26

10 yr. old female caucasian
height = 55 inches, weight 90 lbs.
1984 Summer Haze Study
Fairview Lake YMCA Camp

Figure 2. Plots of pulmonary function data for one child versus maximum one hour ozone concentration, illustrating use of regression to determine slope. (After Spektor et al.[7])

Ozone's transient effects on pulmonary function have been best studied in a series of investigations of children at summer camps in areas with high summer-time ozone levels. The camps are an attractive setting for epidemiologic studies for several reasons. Children are an important group to study, both because the effects of ozone on the developing lung are of intrinsic concern, and because children are less likely to have significant confounding exposures from smoking and occupational hazards. The campers spend most of the day outside, where ambient ozone levels are highest, and are usually quite active, increasing their minute ventilation, and thus their lung dose.

An ozone camp study was conducted in Indiana, Pennsylvania in the summer of 1980. Eighty-three children (31 boys and 52 girls) participated in the study for the two-week day camp session.[8] Pulmonary function testing by spirometer and mini-Wright peak flow meter was performed, with a majority of the children included each day of the program. Pulmonary function measurements included FVC, FEV_1, and PEFR. Peak one-hour ozone levels were estimated by interpolation from surrounding ozone monitoring stations. Peak ozone levels on days of pulmonary function testing ranged from 46 to 110 ppb.

A negative association was seen between level of FVC, FEV_1 and PEFR, and increasing levels of ozone. Peak one-hour ozone level was associated with an estimated mean decrease of FVC by -1.06 mL/ppb and FEV_1 by -0.78 mL/ppb among the 58 children with four or more measurements (Table 1). The distribution of slopes (Figure 3), shows that most children had small negative

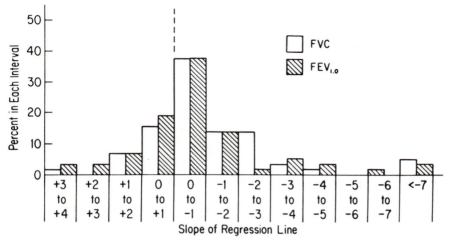

Figure 3. Distribution of the slopes of the individual regressions in mL/ppb of O_3 for the 58 children whose spirometry was measured for 4 or more days in the Indiana, Pennsylvania, camp study.[8]

slopes. In a subsequent paper,[9] peak one-hour ozone was reported to be associated with PEFR by a mean slope of -2.62 mL/sec/ppb among the 23 children with measurements on eight or more days (Table 1) and -1.77 mL/sec/ppb for 33 children measured on six or more days. Response appeared to be lower on rainy days, when children were less active. The authors argue that there could not be an association with particulate air pollution or temperature based on inspection of the data.

A similar study was performed at a day camp in Mendham, New Jersey, in the summer of 1982.[9,10] Sixty-two children between 7 and 13 years of age were tested on 16 days during the five-week camp. Measurements were not made on rainy days. The children in this program were reported to be considerably less active than those in the previous camp. An air pollution episode occurred during the first week of the study, bringing four days of hazy weather and a peak one-hour ozone level of 186 ppb, well above the 120 ppb standard. The mean ozone effects on FVC and FEV_1 were lower than in the previous study, $-.12$ mL/ppb and $-.28$ mL/ppb, respectively, perhaps because of the lower activity levels. Decrements in PEFR, however, were very similar to those seen in the earlier study, -3.0 mL/sec/ppb (Table 1). No significant associations were found with TSP, SO_4, or aerosol acidity. The authors attempted to identify a potentially sensitive subsample based on history of respiratory symptoms, but the numbers were too small to differentiate responses.

Perhaps the most interesting finding of this study was the observation that decrements in lung function persisted for about a week following the haze episode.[10] The authors labeled this prolonged effect as a "persistent" decrement in

Table 1.   Studies of Effects of Ozone on FEV_1 and PEFR.

Authors	Setting	Range of O_3 ppb	Subjects (n)	Effect Estimates	
				FEV_1 (mL/ppb)	PEFR (mL/sec/ppb)
Lippman et al. 1983[8]	Camp, PA	28–122[a]	Boys (24)	− .41	−
			Girls (34)	− 1.04	−
			Total (58)	− .78	− 2.62[b]
Bock et al.[9] Lioy et al. 1985[10]	Camp, NJ	21–186[a]	Boys (17)	− .11	− 1.75
			Girls (22)	− .42	− 3.95
			Total (39)	− .28	− 2.99
Spektor et al. 1987[7]	Camp, NJ	40?–113[a]	Boys (53)	− 1.05	− 7.12
			Girls (38)	− 1.94	− 6.30
			Total (91)	− 1.42	− 6.78
Kinney et al. 1986[14]	Schools, TN	7–78[a]	Children (154)	− .99[c]	−
McDonnell 1985[15]	Chamber Study	120[d]	Boys (22)	− .56	− 1.03
Avol 1985[16]	Chamber Study	150[e]	Children (59)	− .83	− 1.63
Selwyn 1985[17]	Outdoor Running, TX	4–135[f]	Adults (24)	− .40	− 2.38[g]

[a]One hour peak O_3 conc. during camp testing days.
[b]Reported by Bock et al.,[9] 33 children.
[c]$FEV_{.75}$.
[d]Constant 2 1/2 hour exposure.
[e]Constant 1 hour exposure.
[f]Average during running period.
[g]$FEF_{.2-1.2}$.

function, and noted that it complicated the assessment of the "transient" decrements caused by daily ozone peaks.

The third camp study was conducted at a four-week residential program in Fairview, New Jersey, in the summer of 1984.[7] This study was larger than the previous two (91 children), lasted longer, and drew subjects from a high-activity program comparable to the first camp study. Ozone levels were measured at the camp, and had a maximum one-hour concentration of 120 ppb.

Ozone's effects on FEV_1 and PEFR were found to be somewhat larger than in the previous studies: −1.42 mL/ppb and −6.78 mL/sec/ppb, respectively. No associations were found with particle concentrations. The distribution of regression slopes was clearly shifted toward negative values for FEV_1, and even more so for PEFR (Figure 4).

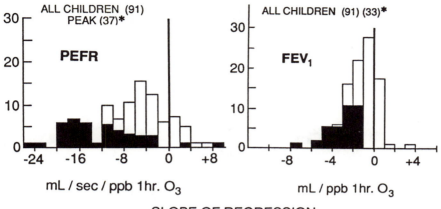

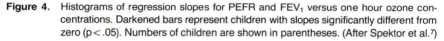

SLOPE OF REGRESSION

Figure 4. Histograms of regression slopes for PEFR and FEV_1 versus one hour ozone concentrations. Darkened bars represent children with slopes significantly different from zero ($p < .05$). Numbers of children are shown in parentheses. (After Spektor et al.[7])

The effect on PEFR in this third camp study is particularly striking, being more than twice the magnitude estimated in the previous studies. Nevertheless, these three studies are in good agreement, and together provide strong evidence that a measurable decrement in function occurs in children exposed to ozone levels below the current NAAQS standard. The studies also suggest that this effect is best measured by decrements in peak flow, and to a lesser extent in FEV_1. These general conclusions are strengthened even further when the results of quite different types of studies, all yielding similar results, are cited.

In a study of families in Tucson, Arizona, daily PEFR measurements were made on children for 11 months using the mini-Wright peak flow meter.[11-13] PEFR varied with both ozone and TSP (Table 2). Neglecting the influence of TSP, ozone levels between 90 and 120 ppb reduced PEFR by approximately 18% when compared to no ozone exposure. (For the children in Mendham, New Jersey, the decrement at 120 ppb was estimated to be approximately 10%.) These results were based on an analysis of variance calculation over all children and days of observation, which underestimates the error of the regression coefficients. This suggests that the results may appear more significant than they would be if the "two-step" analysis used in the camp studies were used. Because the data were presented on a transformed scale, direct comparisons cannot be made with the other studies considered here.

As a part of the ongoing Six Cities Study of the respiratory effects of air pollution, the effects of ozone on pulmonary function were studied in school children in Kingston and Harriman, Tennessee.[14] One hundred and fifty-four children in grades 5 and 6 performed spirometry at each of 6 weekly sessions. Hourly ozone measurements were made at a nearby monitoring station. Ozone levels throughout the study period were quite low, ranging from 7 to 78 ppb.

Table 2.   Average Z Score PEFR in Tucson Children in Relation to Daily Maximum O_3 and Daily Mean TSP [14].

O_3 (ppb)	TSP (μg/m^3)			
	< 56	56–77	77 +	All
< 38	+ .108	+ .239	+ .156	+ .069
38–51	+ .042	− .162	− .061	+ .024
52–79	+ .242	− .021	− .021	− .115
80–120	− .088	− .196	− .804	− .310
All	+ .115	− .027	− .227	

Expressed as standard normal deviates (mean = 0, sd = 1).

Despite the different setting, and substantially lower ozone exposures, the results of this study are in good agreement with the camp studies: decrements in FEV.$_{75}$ of -0.99 mL/ppb and in $\dot{V}_{75\%}$ of -2.40 mL/sec/ppb were estimated (Table 1). No associations were found with fine particulate concentrations.

At least two studies have estimated the effect of ozone on acute decrements in pulmonary function in children under controlled experimental conditions.[15,16] These results are also in approximate agreement with the camp studies. McDonnell et al. exposed 22 boys to 120 ppb ozone for 2.5 hours while exercising vigorously in an environmental chamber.[15] Decrements of $-.56$ mL/ppb FEV$_1$ and -1.03 mL/sec/ppb PEFR were observed (Table 1). Avol et al. used a similar design except that the ozone exposure was provided by bringing the polluted ambient air (144 ppb ozone) into the mobile chamber during a one hour exercise period.[16] During a second hour in the chamber, each subject exercised in filtered air (1 ppb ozone). Decrements in function were $-.80$ mL/ppb for FEV$_1$, and -1.63 mL/sec/ppb for PEFR.

Comparable effects in adults have been investigated in several studies under controlled experimental conditions, but infrequently in the ambient environment. McDonnell and colleagues compared the effects of ozone exposure on pulmonary function in adults and children under the same exposure conditions in an exposure chamber.[15] Intermittent exercise (at a rate producing a constant minute ventilation per unit body surface area) during 2.5 hours of exposure to 120 ppb of ozone produced somewhat larger decrements in PEFR and FEV$_1$ in adults than in children, when measured as a percent of the baseline values (PEFR: -3.9% in adults, -2.7% in children). Adults report more symptomatic complaints, especially cough, from low level ozone exposure than do children similarly exposed.[15,16]

Lebowitz and colleagues have studied peak flow in both adults and children in a household survey in Tucson.[12] After controlling for numerous indoor and outdoor contaminants and weather conditions, an effect of ozone on peak flow was evident in both groups, although the magnitude of this effect is difficult to estimate from their analysis. Selwyn and colleagues studied 24 adults exposed to ambient ozone while running three miles on an outdoor track during summer months in Houston.[17] Ozone levels were measured simultaneously at the site. The average drop in FEV_1 over the approximately 20-minute exercise period was 0.4 mL/sec per ppb of ozone, and this fall was significantly different from zero. Flow between .2 and 1.2 liters ($FEF_{.2-1.2}$) also dropped by -2.4 mL/sec per ppb of ozone.

Transient Effects on Asthma Attacks

Two studies have investigated ozone's effects on attacks of asthma in known asthmatics. Generally, asthmatics are given a diary and asked to record their symptoms for a period of weeks to months. These data are then analyzed in a two-step process. The probability of asthma attack is first estimated for each subject in a logistic regression based on ambient levels of ozone and other factors. Then, as in the pulmonary function analyses, the individual regression coefficients are combined to produce summary effect estimates and to compare subgroups within the sample. (Table 3).

Table 3. Studies of Effects of Ozone on Asthma Attacks in Panel Studies of Adults and Children.

	Study Site	# Subjects	Estimated Relative Risk of 120 ppb Increase in Max Ozone
Whittemore & Korn[18]	Los Angeles	443	1.22
Holguin et al.[19]	Houston	42	2.10

Whittemore and Korn analyzed data from symptoms diaries of 443 asthmatics recruited by the Environmental Protection Agency in the Los Angeles area in 1972–1975.[18] They first proposed the two-step method to analyze these data and pointed out the advantages of estimating individual rather than group attack probabilities.[6] Their method also can account for the correlation among exposure days which occurs because asthmatics are far more likely to have an attack on days immediately following an attack.[6] Whittemore and Korn found that the presence or absence of an attack on the previous day was the strongest predictor of the probability of attack on any given day.[18] After controlling for this auto-correlation, plus day of study, day of week, temperature, relative humidity, and TSP concentration, a small but significant effect of oxidant level (ozone per se was not measured) on asthma attack risk was observed. An increase in oxidant level from 0 to 120 ppb was associated with an estimated relative risk of an attack of 1.22. No differences in response between sexes or by age were observed. The estimated relative risk of attack for a TSP concentration of 760 mg/m³, the 24-hour standard, was 1.23. While there was substantial variability in the individual ozone coefficients consistent with a difference in response between individuals, there was no evidence for TSP estimates to vary among panelists compared to their estimated standard errors.

Holguin and colleagues performed a similar study in Houston in 1981.[19] The study population was smaller, 42 subjects who each had five or more asthma attacks, but compliance with the reporting protocol was more complete, and individual exposures may have been more accurately estimated than in the Whittemore and Korn study. Ambient ozone concentrations had a range of 17 ppb to 265 ppb.[20] The same statistical model for individual attack risk was used. After adjusting for previous attacks, ambient temperature and humidity, pollen counts and nitrogen dioxide level, there was a significant increase in risk of asthma attack associated with ozone level. An increase of 120 ppb in the ozone level was associated with an estimated relative risk of asthma attack of 2.10. The risk of ozone-related asthma attack was slightly higher in females than in males, in those over 18 years versus those younger, and was also higher for those asthmatics living in households without smokers compared to those with smokers. The modifying effects of gender and age were not large enough to be statistically significant in this small study population, but they are noteworthy because of their correspondence to the findings of other studies.

Hospital Admission Studies

Bates and Sizto have also studied the effects of ozone on rates of respiratory disease by correlating hospital admissions data with air pollution levels.[21,22] Data were gathered from the 79 hospitals in southern Ontario that admitted cases of acute respiratory disease for the years 1974–1982. The numbers of daily admissions in nine separate ICD classes of acute respiratory conditions (e.g., asthma, viral pneumonia, acute bronchitis) plus a comparison group of nonrespiratory conditions, were tabulated for 2 winter months and 2 summer months in each year, and correlated with the daily levels of ozone, sulfur dioxide, nitrogen dioxide, and other pollution measures collected by the provincial network of 17 monitoring stations. In a check on the overall validity of the approach, Bates and Sizto showed that admissions for nonrespiratory complaints were not associated with the air pollution levels. Hospital admissions for all respiratory conditions correlated with ozone levels in summer, but not in winter months (when ozone levels were less than half as high). The correlation was strengthened by lagging the exposure data 48 hours behind the admissions data—presumably because the ozone-induced attack required some hours to develop. The association was present for asthma and for all respiratory conditions less asthma. However these associations with ozone were no longer statistically significant when the effects of SO_4, NO_2, and temperature are also taken into account. Bates and Sizto concluded that sulfates were the more important determinant of respiratory hospital admissions, although both O_3 and SO_4 may be surrogates for an unknown constituent of summertime polluted air masses.

DISCUSSION

Epidemiologic Study Design

The study of infrequent, reversible respiratory health changes associated with short-term exposures to air pollutants in the general environment has, until recent years, been intractable to the environmental epidemiologist. Precise daily incidence data which could be correlated with measures of daily pollution were almost impossible to determine. The only defined endpoint with good daily incidence information is total mortality. Several studies have attempted to relate daily total mortality to concurrent air pollution concentrations. While mortality data for respiratory diseases can be gleaned from death certificates, the poor quality of the diagnoses and the wide range of factors contributing to each cause of death seriously limit attempts to study air pollution mortality.

Respiratory morbidity incidence is even harder to define because of the lack of a consistent definition of the respiratory endpoint. Diagnosis of pulmonary disease may vary between doctors and between different types of health care institutions. Even assuming a well-defined diagnosis, the definition of the population from which the cases have arisen is difficult. The Bates and Sizto papers[21,22] are unique in being able to classify all respiratory admissions in all acute care hospitals in a large region with an air pollution monitoring network. Daily incidence data can be regressed directly against daily air pollution and meteorologic measures. Hospital admission studies are a type of "ecological study" in which broad indices of effect are correlated with average measures of exposures. These studies are limited in usefulness because of the difficulty of deriving quantified estimates of the magnitude of an air pollution effect from such aggregate data.

In general, studies of short-term respiratory effects are based on panels of a few score to a few hundred individuals who periodically report their symptoms or who have repeated measures of pulmonary function. The panels are often made up of subjects thought to be "susceptible" to the effects of air pollution, to increase the statistical power of the experiment by raising the background incidence or prevalence rates. Unfortunately, these panel studies are subject to potential bias associated with the correlation of response between successive days, and the potential bias due to "drop-ins" and "drop-outs" of panelists with high or low attack frequencies.

In 1979, Korn and Whittemore[6] showed how data could be collected and analyzed in such panel studies to avoid these potential biases. The method is based on a two-step process. In the first step, the data for each individual is regressed on air pollution and other covariates. In the second step, the individual air pollution regression coefficients (effect estimates) are analyzed to determine mean effect estimates and characteristics of the variability of individual effect estimates within the population. Whittemore and Korn[18] applied the method to the analysis of asthma attacks and oxidant exposures in Los Angeles. Holguin and co-workers have applied the same methods to similar data from Houston.[19] The method was subsequently applied to the analysis of pulmonary function data in the summer camp and related studies of pulmonary function and ozone.

Thus, the development of methods for collecting and analyzing panel data has been followed by a series of epidemiologic studies of the associations of short-term ozone exposures with increased asthma attacks and reduced pulmonary function.

Chamber studies of ozone's effects on volunteer subjects also have made important contributions to the design of epidemiologic studies of ozone's short-term effects. In particular, the role of exercise in ozone-induced airway narrowing has been well-studied in chambers, and this has led to careful consideration of exercise levels in designing epidemiologic studies.

Air Pollution Exposures

Historically, epidemiologic studies of the health effects of air pollution have examined populations in large metropolitan areas or in areas with substantial emissions of pollutants from local industries, in order to find populations with known, predictable exposures. As the monitoring network has expanded, it has become clear that secondary pollutants, such as ozone, are found in high concentrations throughout the United States, in rural as well as urban areas. In fact, primary pollutants in urban air masses initially scavenge ozone so that concentrations are often lower in cities than for adjacent areas. Addition of these primary pollutants ultimately leads to higher concentrations of photochemical oxidants and other secondary pollutants as the air mass moves downwind.

Recent epidemiologic studies have sampled subjects in rural areas downwind from important source regions for primary pollutants. The difficulty with this strategy is that observing a high pollution event requires being at the right place at the right time. In addition to a source of pollution, the wind trajectories must carry the air mass over the study site, and sunlight, temperature, and other factors must be "optimal" to allow the formation of photochemical oxidants. Thus, a successful acute study in a population monitored only a few weeks requires not only careful planning, but a reasonable amount of good luck.

Ozone is a secondary pollutant with highest exposures often far downwind from source regions. The same conditions which produce photochemical oxidants are also likely to promote the production of other secondary pollutants such as sulfuric and nitric acid, their neutralized species (sulfates and nitrates), and other fine particles. Ozone events are characterized by hot, *hazy* weather, with the hazy conditions produced by the suspended fine particles and droplets in the air mass. Thus, high ozone exposures are likely to be accompanied by high concentrations of other pollutants which have also been shown to produce transient declines in lung function. Several of these studies have attempted to differentiate the effects of ozone and various particle species. However, the limited range of exposure data for the various constituents limits their ability to separate the effects of these pollutants, and no firm conclusions about differential effects can be made.

Nevertheless, the problem of mixed exposures can be put in perspective by noting that controlled laboratory experiments exposing small numbers of selected subjects to ozone in filtered air without other pollutants[15] have shown deficits of lung function of similar magnitude to those observed in populations exposed to ozone in outdoor air. Thus, while the other pollutants cannot be ignored, ozone must be an important independent cause of short-term respiratory effects.

Misclassification of Response

The two-step analysis we have described in which individual effect estimates are first calculated and then pooled, allows us to examine the distribution of regressions of pulmonary function and asthma attack rates on ozone concentration. These distributions have shown a shift in the response with ozone exposure, and have illustrated the range of the estimated response. There is the temptation to consider subjects in one tail of the distribution as being more susceptible. This is an incorrect interpretation of the data. Considerable intra-subject variability exists in these data, and an individual's position within the group distribution is subject to considerable error. It is the overall shift in the distribution that is revealing, not who happens to fall in the tails.

On the other hand, subgroups with the potential to be more susceptible may be defined by ''independent'' risk factors, such as history of asthma. Bock et al.[9] tested for such differences in the Indiana, Pennsylvania, camp study, and although the children with a history of respiratory illness appeared to have greater response, the numbers of subjects were too small for statistical significance. Future studies may be designed to oversample such subjects, who are defined a priori as being potentially more susceptible.

Dose Modeling

There are several ways that our understanding of ozone's effects on the lungs can be improved. One of the most important of these is by more accurately estimating the true lung dose of each of the individual subjects under study. Inaccurate estimation of lung dose leads to misclassification of exposure status, and hence to an underestimation of the magnitude of the effect being studied. Attention is often given to accurate measurement of ambient ozone in the vicinity of the subjects of an epidemiologic study, but this is only the first step in dose estimation. The actual tissue concentration of ozone in the lung and the time course of changes in that concentration are only partially determined by the inhaled concentration. Other determinants are an individual's ventilation rate (in turn affected by physical exertion, age, training, etc.), nose versus mouth breathing, and a variety of different variable attributes of the airways such as permeability.

Mage and colleagues[23] have calculated the ozone dose to the lungs (mass of ozone delivered to the trachea) for children in a Canadian camp study. Their calculations consider the activity levels of the children and the amount of time spent indoors, when ozone levels are lower than measured at outdoor monitors. Their calculations confirm that outdoor ozone concentration is a poor surrogate for ozone lung dose. For example, an increase in activity resulting in a one standard deviation rise in heart rate can raise estimated lung dose by 70%.

Miller and colleagues have constructed several compartmental models for estimating the regional deposition of ozone in the respiratory tract.[24] These models basically begin where Mage leaves off—at the trachea, and estimate the fraction of the ozone that reaches tissue in the airways and lung parenchyma. As an example of what can be learned from such a model, the authors calculated that increased ventilation would increase uptake of ozone in the pulmonary tissue, but affect airway tissue levels relatively little.

Yet even Miller's detailed compartmental model does not provide guidance on the time course of ozone's effect on the lungs. Most studies have attempted to associate changes in respiratory status with maximum one-hour ozone concentrations. A simple example will illustrate the difficulty with this approach. Suppose that we are studying ozone's effects on pulmonary function in children at two camps, and are comparing each child's daily PEFR (measured at the end of the day) to the day's maximum hourly ozone concentration. Suppose, further, that at one camp the day's maximum ozone level occurs late in the afternoon, i.e., shortly before the PEFR measurement, while at the other camp the peak occurs early in the morning. All other things being equal, are the data from the two camps comparable? Is the day's maximum concentration a valid measure of ozone exposure, regardless of how long before the PEFR measurement that peak occurred? Clearly the answer depends on the time course of ozone's effects on the lung. How closely in time do changes in the functional response follow changes in exposure? Are there nonlinear kinetics such as an enhanced response at high dose rates that may complicate the relationship between lung dose and response?

Some authors have taken a statistical approach to this problem by fitting different exposure variables—the maximum ozone level, the daily cumulative exposure, etc., to the response data and asking which variable seemed to fit better.[7] This approach is of limited usefulness because epidemiologic data are generally too weak to distinguish amongst competing models.[25] It would seem preferable to derive a model from biological principles rather than relying on the one that happens to fit somewhat better the data in hand. If this seems arbitrary it may be because we are not accustomed to mixing epidemiologic and toxicologic models. The advantage of this approach is that an exposure-response model whose form is suggested by pathophysiology rather than by a variable selection routine in a computer regression package can be supported or refuted by properly designed experiments: chamber studies with humans, animal studies, etc.

Smith has proposed just such an approach to exposure assessment, and illustrated it with a simple two-compartment pharmacodynamic model for ozone in the lungs.[26] In Smith's model, changes in effective dose lag about one hour behind changes in air concentration, and exposures more than a few hours before pulmonary function measurement have very little impact on the effective dose unless they are of very high intensity. Kriebel and Smith[27] have recently proposed a more refined four compartment version of this model.

There are practical difficulties in using these models. They are inevitably based on an incomplete understanding of the toxic mechanisms, and the available data are often inadequate. Nevertheless, this approach seems a rational method for deciding how exposure data should be used in epidemiologic analyses. Furthermore, this research strategy suggests a clear and vital link between toxicology, environmental monitoring, and epidemiology: the development of effective dose models for epidemiology identifies new and immediately useful questions for toxicologists and air pollution scientists to answer, and the results of the epidemiologic analyses using the effective dose estimates provide confirmation of the models themselves and the toxicologic principles underlying them.

CONCLUSIONS

Recent advances in the epidemiologic assessment of the acute health effects of ozone exposures can be attributed, at least in part, to changes in the design of these studies based on (a) the temporal and spatial distribution of ozone, (b) the development of new analytic methods, and (c) the consideration of endpoints and modifying factors shown to be important in controlled exposure studies of humans. Further advances could be expected with a better understanding of the dose of ozone delivered to the lungs. In particular, repeated pulmonary function tests during and following controlled exposures, to characterize the time course of the loss and recovery of pulmonary function, would allow estimation of the appropriate time constants for measuring effective dose in epidemiologic studies.

ACKNOWLEDGMENTS

This study was supported in part by NIEHS Grant ES 0002. Dr. Kriebel was supported in part by NIEHS Grant ES 04202, and Dr. Dockery was supported in part by NIEHS Grant ES 01108.

REFERENCES

1. "Air Quality Criteria for Ozone and Other Photochemical Oxidants," U.S. EPA Report EPA/600/8-84/020eF, Volume V. (1986).
2. Detels, R., S. N. Rokaw, and A. H. Coulson, et al. "The UCLA Population Studies of Chronic Obstructive Respiratory Disease: I. Methodology and Comparison of

Lung Function in Areas of High and Low Pollution," *Am. J. Epidemiol.* 109:33–58 (1979).

3. Rokaw, S. N., R. Detels, and A. H. Coulson, et al. "The UCLA Population Studies of Chronic Obstructive Respiratory Disease: 3. Comparison of Pulmonary Function in Three Communities Exposed to Photochemical Oxidants, Multiple Primary Pollutants, or Minimal Pollutants," *Chest* 78:252–262 (1980).

4. Detels, R., J. W. Sare, and A. H. Coulson, et al. "The UCLA Population Studies of Chronic Obstructive Respiratory Disease: IV. Respiratory Effect of Long-Term Exposure to Photochemical Oxidants, Nitrogen Dioxide, and Sulfates on Current and Never Smokers," *Am. Rev. Resp. Dis.* 124:673–680 (1981).

5. Kulle, T. J., L. R. Sauder, S. J. Hebel, and M. D. Chatham. "Ozone Response Relationships in Healthy Nonsmokers," *Am. Rev. Resp. Dis.* 132:36–41 (1985).

6. Korn, E. L., and A. S. Whittemore. "Methods for Analyzing Panel Studies of Acute Health Effects of Air Pollution," *Biometrics* 35:795–802 (1979).

7. Spektor, D. M., M. Lippmann, P. J. Lioy, G. D. Thurston, K. Citak, N. Bock, F. E. Speizer, and C. Hayes. "Effects of Ambient Ozone on Respiratory Function in Active Normal Children," *Am. Rev. Resp. Dis.* 137:313–320 (1988).

8. Lippmann, M., P. J. Lioy, G. Leikauf, K. B. Green, D. Baxter, M. Morandi, B. S. Pasternack, D. Fife, and F. E. Speizer. "Effects of Ozone on the Pulmonary Function of Children," in S. D. Lee, M. G. Mustafa, and M. A. Mehlman, eds. *International Symposium on the Biomedical Effects of Ozone and Related Photochemical Oxidants* , March, 1982, Pinehurst, NC. (Princeton NJ: Princeton Scientific Publishers, Inc.; Advances in Modern Toxicology: v. 5), pp. 423–446.

9. Bock, N., M. Lippmann, P. Lioy, A. Munoz, and F. E. Speizer. "The Effects of Ozone on the Pulmonary Function of Children," in S. D. Lee, ed. *Evaluation of the Scientific Basis for Ozone/Oxidants Standards*, November, 1984, (Houston, TX: Air Pollution Control Association TR4), pp. 297–308.

10. Lioy, P. J., T. A. Vollmuth, and M. Lippmann. "Persistence of Peak Flow Decrements in Children Following Ozone Exposures Exceeding the National Ambient Air Quality Standard," *J. Air Pollut. Contr. Assoc.* 35:1068–1071 (1985).

11. Lebowitz, M. D., C. J. Holberg, and R. R. Dodge. "Respiratory Effects on Populations from Low-Level Exposures to Ozone," Paper 83-12.5. Annual Meeting of Air Pollution Control Association, Atlanta, GA, June 1983.

12. Lebowitz, M. D., C. J. Holberg, B. Boyer, and C. Hayes. "Respiratory Symptoms and Peak Flow Associated with Indoor and Outdoor Air Pollutants in the Southwest," *J. Air Polut. Contr. Assoc.* 35:1154–1158 (1985).

13. Lebowitz, M. D. "The Effects of Environmental Tobacco Smoke Exposure and Gas Stoves on Daily Peak Flow Rates in Asthmatic and Non-Asthmatic Families," *Europ. J. Resp. Dis.*, Suppl 133, 65:90–97 (1984).

14. Kinney, P. L. "Acute Lung Function Change in Children Exposed to Community Air Pollution," Dissertation, Department of Environmental Science and Physiology, Harvard School of Public Health, November 1986.

15. McDonnell, W. F., R. S. Chapman, D. H. Horstman, M. W. Leigh, and S. Abdul-Salaam. "A Comparison of the Responses of Children and Adults to Acute Ozone Exposure," in S. D. Lee, ed. *Evaluation of the Scientific Basis for Ozone/Oxidants Standards*, November 1984 (Houston, TX: Air Pollution Control Association, TR-4:317-328, 1985).

16. Avol, E. L., W. S. Linn, D. A. Shamoo, L. M. Valencia, U. T. Anzar, T. G. Venet, and J. D. Hackney. "Respiratory Effects of Photochemical Oxidant Air Pollution in Exercising Adolescents," *Amer. Rev. Respiratory Dis.* 132:619–622 (1985).

17. Selwyn, B. J., T. H. Stock, R. J. Hardy, F. A. Chan, D. E. Jenkins, D. J. Kotchmar and R. S. Chapman. "Health Effects of Ambient Ozone Exposure in Vigorously Exercising Adults," in S. D. Lee, ed. *Evaluation of the Scientific Basis for Ozone/Oxidants Standards*, November 1984 (Houston, TX: Air Pollution Control Association TR-4:281–296).

18. Whittemore, A. S., and E. L. Korn. "Asthma and Air Pollution in the Los Angeles Area," *Amer. J. Public Health* 70:687–696 (1980).

19. Holguin, A. H., P. A. Buffler, C. F. Contant, T. H. Stock, D. Kotchmar, B. P. Hsi, D. E. Jenkins, B. M. Gehan, L. M. Noel, and M. Mei. "The Effects of Ozone on Asthmatics in the Houston Area," in S. D. Lee, ed. *Evaluation of the Scientific Basis for Ozone/Oxidants Standards*, November 1984 (Houston, TX: Air Pollution Control Association TR-4:262–280).

20. Stock, T. H., D. J. Kotchmar, C. F. Contant, et al. "The Estimation of Personal Exposures to Air Pollutants for a Community-Based Study of Health Effects in Asthmatics," *J. Air Pollut. Contr. Assoc.* 35:1266–1273 (1985).

21. Bates, D. V., and R. Sizto. "Relationship Between Air Pollution Levels and Hospital Admissions in Southern Ontario," *Canadian J. Public Health* 74:117–122 (1983).

22. Bates, D. V., and R. A. Sizto. "A Study of Hospital Admissions and Air Pollutants in Southern Ontario," Chapter 58 in *Aerosols* S. D. Lee, T. Schneider, L. D. Grant, and P. J. Verkerk eds. (Chelsea, MI: Lewis Publishers, 1986).

23. Mage, D. T., M. Raizenne, and J. Spengler. "The Assessment of Individual Human Exposures to Ozone in a Health study," in S. D. Lee, ed. *Evaluation of the Scientific Basis for Ozone/Oxidants Standards*, November 1984 (Houston, TX: Air Pollution Control Association TR4), pp. 238–249.

24. Miller, F. J., J. H. Overton, R. H. Jaskot, and D. B. Menzel. "A Model of the Regional Uptake of Gaseous Pollutants in the Lung," *Tox. and Applied Pharm.* 79:11–27 (1985).

25. Robins, J. M., and S. Greenland. "The Role of Model Selection in Causal Inference from Nonexperimental Data," *Amer. J. Epidemiology* 123:392–402 (1986).

26. Smith, T. J. "Exposure Assessment for Occupational Epidemiology," *Am. J. Ind. Med.* 12:249–268 (1987).

27. Kriebel, D., and T. J. Smith. "A New-Linear Pharmacologic Model of the Acute Effects of Ozone on the Human Lung," *Env. Res.* (In press, 1990).

CHAPTER 11

Basis for the Primary Ozone Standard

David J. McKee

INTRODUCTION

National ambient air quality standards (NAAQS) for photochemical oxidants (ozone) were promulgated (36 FR 8186) on April 30, 1971 to protect public health and welfare from adverse effects of ambient oxidants. Both standards were set at an hourly average of 0.08 parts per million (ppm) not to be exceeded more than 1 hour per year. Subsequent review of criteria resulted in promulgation of the current standard (44 FR 8202) on February 8, 1979 which (a) raised the primary (health-based) standard to 0.12 ppm, (b) raised the secondary (welfare-based) standard to 0.12 ppm, (c) changed the chemical designation of the standard from photochemical oxidants to ozone, (d) changed to a standard with a statistical rather than deterministic form, and (e) changed the definition of the point at which the standard is attained to ''when the expected number of days per calendar year with maximum hourly average concentrations above 0.12 ppm is equal to or less than one.''

NAAQS REVIEW PROCESS

The NAAQS review process has become somewhat complex and lengthy. The process involves numerous individuals and organizations, both inside and outside of the Environmental Protection Agency (EPA). Many analyses and documents, which are reviewed within EPA and by the public, have become a necessary part of the total process of NAAQS review. A summary of key steps in development of the NAAQS is presented in Figure 1. More detailed discussion of this process is available in other documents.[1-3]

Development of the criteria document is the first step. This document is a comprehensive review of the scientific literature upon which a particular NAAQS is based. After undergoing an extensive review by the public and the Clean Air Scientific Advisory Committee (CASAC), appropriate revisions are made by the Environmental Criteria and Assessment Office (ECAO) until "final closure" on the document is received from CASAC. As the criteria document nears completion, staff of the Office of Air Quality Planning and Standards (OAQPS) begin writing a "staff paper" which critically assesses scientific information contained in the criteria document from a standard-setting perspective. The staff paper also summarizes and discusses the implications of various analyses (e.g., air quality, exposure, risk) for standard-setting. The staff paper, which is also reviewed by the public and CASAC, serves as the basis for staff recommendations on the NAAQS to the Administrator. Recent staff papers have included ranges for standards.

The staff recommendations are further developed in a decision package which is reviewed extensively within EPA. After incorporation of comments, the package is sent to the Administrator for a decision. This decision is formalized by publication in the *Federal Register*. If the decision is to propose a NAAQS or changes in a NAAQS, the notice informs the public of the proposed standard(s) and requests public comments. Following a comment period and public meetings, the EPA reviews and incorporates comments to the package as appropriate. Internal EPA review is continued until the Administrator makes a final decision and the NAAQS is promulgated in the *Federal Register*.

The ozone NAAQS currently are under review. The ozone criteria document[4] has been reviewed by the public and CASAC, and a final draft was published in October 1986. The staff paper[5] also has been reviewed by the public and has received closure from CASAC. Air quality, exposure, and risk analyses will be completed and given consideration prior to development of a regulatory decision package and subsequent proposal.

HEALTH EFFECTS OF OZONE

Evidence for health effects from exposure to ozone (O_3) has come from both human and animal studies. The strongest and most quantifiable data are provided by controlled human exposure studies, but these studies are limited to acute (short-term) exposures and noninvasive techniques. Field studies are similarly limited, but permit investigation of the effects of oxidants in ambient air and allow for better characterization of exposure than epidemiological studies. Although use of most community studies has been hampered by the difficulty in adequately

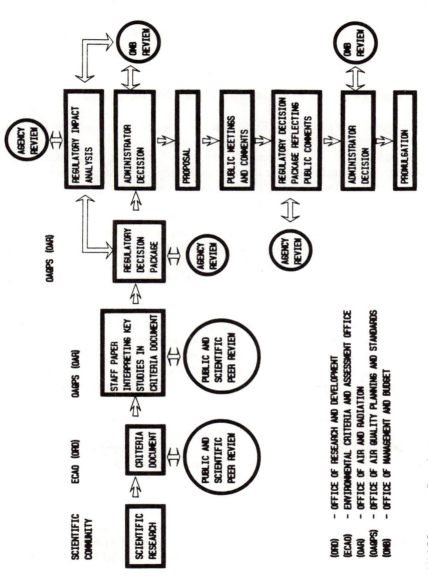

Figure 1. NAAQS process flow chart.

SCIENTIFIC COMMUNITY

ECAO (ORD)

OAQPS (OAR)

OAQPS (OAR)

SCIENTIFIC RESEARCH

CRITERIA DOCUMENT

PUBLIC AND SCIENTIFIC PEER REVIEW

STAFF PAPER INTERPRETING KEY STUDIES IN CRITERIA DOCUMENT

PUBLIC AND SCIENTIFIC PEER REVIEW

REGULATORY DECISION PACKAGE

AGENCY REVIEW

AGENCY REVIEW

REGULATORY IMPACT ANALYSIS

AGENCY REVIEW

OMB REVIEW

ADMINISTRATOR DECISION

PROPOSAL

PUBLIC MEETINGS AND COMMENTS

REGULATORY DECISION PACKAGE REFLECTING PUBLIC COMMENTS

ADMINISTRATOR DECISION

OMB REVIEW

PROMULGATION

(ORD)   – OFFICE OF RESEARCH AND DEVELOPMENT
(ECAO)   – ENVIRONMENTAL CRITERIA AND ASSESSMENT OFFICE
(OAR)   – OFFICE OF AIR AND RADIATION
(OAQPS) – OFFICE OF AIR QUALITY PLANNING AND STANDARDS
(OMB)   – OFFICE OF MANAGEMENT AND BUDGET

characterizing exposure and numerous confounding variables, these investigations provide important supporting evidence for effects occurring in populations. Animal toxicology studies have provided evidence of acute and chronic (long-term) exposure effects which can be detected only with invasive procedures, but uncertainties inherent in dosimetry and species sensitivity differences have limited quantitative extrapolation to humans.

Assessment of health effects attributed to O_3 requires consideration of the database in each of the above areas of study. This section integrates research in each of these areas to provide an indication of the strength of the database for each effect.

Alterations in Pulmonary Function

The best documented and strongest evidence of human health effects of O_3 exposure are pulmonary function decrements. Controlled human exposure, field, epidemiology, and animal toxicology studies have provided evidence that exposure to O_3 can modify such pulmonary measurements as forced expiratory volume (FEV), forced expiratory flow (FEF), forced vital capacity (FVC), functional residual capacity (FRC), vital capacity (VC), tidal volume (V_T), peak expiratory flow rate (PEFR), inspiratory capacity (IC), residual volume (RV), total lung capacity (TLC), airway resistance (R_{aw}) and breathing frequency (f_B).

Early controlled experimental studies of resting human subjects exposed to O_3 levels up to 0.75 ppm for 2 hours demonstrated little or no change in FVC,[6,7] FEV_1, and FRC.[6] Flow rate variables such as FEF 25% and FEF 50% showed up to 30% decreases in some subjects exposed at rest to 0.75 ppm O_3,[6,8] while only small increases in R_{aw} (<17%) were reported for >0.5 ppm O_3 exposures.[8,9] More recent studies have reported occurrence of FEV and FEF decrements during resting exposures to ≥0.5 ppm O_3;[10,11] however, no statistically significant changes in R_{aw} and only suggestive changes in RV and TLC have been reported for similar exposures.[12] Airway resistance is not generally affected in resting subjects at these O_3 levels. Changes in pulmonary function do not occur in resting subjects exposed to ≤0.3 ppm O_3,[10] though some subjects exhibit O_3-induced pulmonary symptoms during resting exposures.[9,13] In general, however, because subjects were at rest in most of the older studies, significant respiratory effects were not reported even for higher O_3 exposures.

Exercise, which causes increased minute ventilation (V_E), enhances individual and group mean response to O_3 exposure. Exercise increases breathing frequency and depth of breathing resulting in greater total dose of O_3 inhaled and increased penetration to the most sensitive lung tissue. As exercise levels increase to the point where ventilatory rate (V_E) exceeds approximately 35 L/min, oronasal or oral breathing tends to predominate;[14] thus at higher exercise levels a greater portion of the inhaled O_3 will bypass the nose and nasopharynx.[15] Individual variability will affect the V_E at which oral or oronasal breathing predominates. This further increases the total dose of O_3 to the lower airways and parenchyma. The relationship between exercise and magnitude of response is illustrated quite well by Figure 2, prepared for the criteria document[4], showing group mean decrements in FEV_1 caused by exposure of exercising subjects to various O_3 levels based on results from 25 different studies. The curves clearly demonstrate that as exercise levels increase for a given O_3 exposure, there is a resultant increase in the group mean FEV_1 decrements. These curves also provide estimates of group mean FEV_1 decrements resulting from exposure of subjects to O_3 under light, moderate, heavy, and very heavy exercise conditions. Individual curves presented in the criteria document,[4] suggest an increase in subject response variability as exercise levels increase.

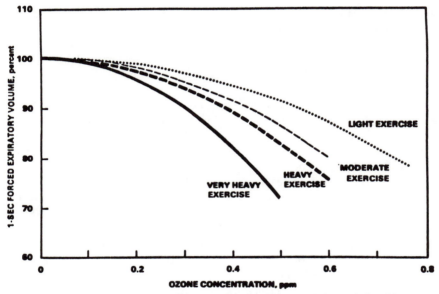

Figure 2. Group mean decrements in 1-sec forced expiratory volume during 2-hr ozone exposures with different levels of intermittent exercise: light ($V_E < 23$ L/min); moderate ($V_E = 23$–43 L/min); heavy ($V_E = 44$–63 L/min); and very heavy ($V_E > 64$ L/min). Concentration-response curves are taken from Figures 13-2 through 13-5 of the ozone criteria document.[4]

Pulmonary function decrements have been reported in controlled exposure studies of healthy exercising human subjects exposed to O_3 levels in the range of 0.12 to 0.18 ppm. During 2 hours of intermittent very heavy exercise ($V_E = 64$ L/min.), healthy subjects experienced group mean decrements in FEV_1 of 4.5, 6.4, and 14.4% for O_3 exposures of 0.12, 0.18, and 0.24 ppm, respectively, with greater FEV_1 decrements at higher O_3 concentrations.[16] Positive associations between group mean magnitude of FVC (3, 4, and 12%) and FEF (7.2, 12, and

23%) decrements and O_3 exposures of (0.12, 0.18, and 0.24 ppm) also were reported in the McDonnell et al. study;[16] statistically significant increases in SR_{aw} and f_B did not occur until O_3 was ≥ 0.24 ppm. In a separate study, McDonnell et al.[17] reported a statistically significant but small decline (-3.4%) in FEV_1 of children (8–11 years) after 2 hours of exposure to 0.12 ppm O_3 during intermittent heavy exercise ($V_E = 39$ L/min). Furthermore, these small decrements in FEV_1 persisted for 16 to 20 hours after O_3 exposure ceased.

Support for O_3-induced pulmonary function changes also comes from other controlled exposure studies. Avol et al.[18] report statistically significant, small decreases (6.1%) in FEV_1, at 0.16 ppm O_3 with larger decrements at ≥ 0.24 ppm O_3. Decrements are reported in FEV_1, FVC, and FEF for distance runners and distance cyclists exposed to 0.20 ppm and 0.21 ppm O_3 respectively, during 1 hour of continuous, heavy exercise ($V_E = 77.5$ and 81 L/min).[19,20] In another study involving O_3 concentrations ranging between 0.10 and 0.25 ppm, exponential decreases in FVC, FEV_1, FEF_{25-75}, SG_{aw}, IC, and TLC have been reported with exposure to increasing O_3 concentration during very heavy exercise ($V_E = 68$ L/min); time of exposure was related to linear decreases in FVC and FEV_1.[21]

Field studies, which contain elements of both controlled human exposure and epidemiologic studies, provide the most quantitatively useful human exposure data available for ambient photochemical oxidants. Results from field studies are consistent with pulmonary function decrements reported in controlled O_3 exposure studies. For a 1-hour exposure to mean O_3 concentrations of 0.144 ppm, small significant group mean decreases in FVC (-3.3%), $FEV_{0.75}$ (-4.0%), FEV_1 (-4.2%), MMFR (-3.2%) and PEFR (-3.9%) relative to pre-exposure levels are reported in 59 healthy continuously exercising ($V_E = 32$ L/min) adolescents.[22,23] In a separate study of 50 healthy, adult, continuously exercising ($V_E = 53$ L/min) bicyclists, mean O_3 concentrations of 0.153 ppm for 1 hour produce statistically significant group mean decreases in FEV_1 (-5.3%) compared to pre-exposure.[18] This study[18] showed that similar effects result in subjects exposed to comparable O_3 levels in ambient and controlled O_3 exposures. Small but statistically significant decreases in FEV_1 and FVC were also reported in exercising healthy and asthmatic adults during exposure to mean O_3 concentrations of 0.165 and 0.174 ppm.[24-26] Of importance is the fact that no recovery occurred in healthy adults during a 1 hour post-exposure rest period and FEV_1 remained low or decreased further 3 hours after exposure for asthmatics.[27] Table 1 provides a summary of group mean percent changes in FEV_1 for controlled exposure and field studies.

In summary, pulmonary function decrements have been reported in healthy adult subjects (18 to 45 years old) after 1 to 3 hours of exposure as a function of the level of exercise performed and the O_3 concentration inhaled during exposure. Group mean data pooled from numerous controlled human exposure and field studies and summarized in the criteria document[4] indicate that, on average, group mean pulmonary function decrements occur at:

(a) ≥ 0.5 ppm O_3 when at rest (sitting)
(b) ≥ 0.37 ppm O_3 with light exercise (slow walking)
(c) ≥ 0.30 ppm O_3 with moderate exercise (brisk walking)
(d) ≥ 0.24 ppm O_3 with heavy exercise (easy running)
(e) ≥ 0.18 ppm O_3 with very heavy exercise (competitive running)

However, data from a field study[22,23] indicate that group mean pulmonary function decrements occur during heavy exercise in healthy adults exposed to mean O_3 levels of 0.153 ppm and may occur in very heavily exercising adults during controlled exposures to 0.12 ppm O_3.[16] Group mean pulmonary function decrements have also been reported for children exposed to controlled O_3 levels of 0.12 ppm and adolescents exposed to mean ambient O_3 levels of 0.144 ppm during heavy exercise.[17,22,23] In general, the group mean changes in lung function in the above studies are regarded as small ($<5\%$), but there is considerable intersubject variability in the magnitude of individual pulmonary response. Based on an analysis of the individual data from one controlled human exposure study,[16] approximately 15% of healthy, heavily exercising individuals exposed to 0.12 ppm for 2 hours experience at least a 10% decline in FEV_1.[30] At 0.24 ppm, analysis of individual data from both the McDonnell et al.[16] and Avol et al.[18] studies show about 40% of the healthy exercising subjects with a decline of at least 10% in FEV_1. Thus, it appears that some individuals are intrinsically more responsive, and this group which has been referred to as "responders," may constitute as much as 5% to 20% of the healthy subjects studied.

Further confirmation of the relationship between acute O_3 exposure and pulmonary function decrements is provided by several epidemiological studies of children and young adults.[31-38] These studies report decreased peak flow or increased airway resistance for acute exposures to ambient O_3 concentrations ranging from 0.01 to 0.186 ppm over the entire study period. Of particular importance is the finding by Lioy et al.[38] that a persistent decrement in lung function of children lasted as much as a week after the end of a smog period of four days, during which peak 1-hr O_3 levels were in the range of 0.135 to 0.186 ppm. Whether or not the underlying causes of the persistent decrements suggested by Lioy et al.[38] (i.e., altered epithelial permeability and changes in airway secretion) are confirmed, the decrements represent a potentially more serious response than the more transient effects found in controlled exposure studies.

While it is reasonable to conclude that none of these epidemiological studies provide definitive quantitative data by themselves, due to methodological problems and confounding variables, the aggregation of studies provides reasonably good qualitative evidence of association between ambient O_3 exposure and acute pulmonary effects. The association is strengthened by the consistency of these epidemiological results with the findings of McDonnell et al.[17] and Avol et al.,[22,23] who reported small decrements in pulmonary function for exercising children exposed to 0.12 ppm O_3 in purified air, and adolescents exposed to 0.144 ppm O_3 in ambient air, respectively.

Table 1. Lung Function Changes and Symptoms Attributed to Ozone Exposures.

Ozone Concentration ppm	Measurement[a,b] Method	Exposure Duration	Activity[c] Level (V_E)	Group Mean % Change in FEV_1 (Range for O_3 Exp.)	Symptoms Reported	No., Sex and Age of Subjects	Reference
0.00 0.08	UV,UV	1 hr	CE (57)	+0.6 +1.7 (ns)	None	42 male 8 female (26.4 ± 6.9)	Avol et al.[18]
0.00 0.10	UV,UV	2 hr	IE (68)	+1.3 +1.1 (nd) (range: +/0 to −10)	None	20 male (25.3 ± 4.1 yr)	Kulle et al.[21]
0.00 0.10	CHEM,NBKI	2 hr	IE (67)	−0.3 −2.6 (ns)	None	10 male (18–28 yr)	Folinsbee et al.[10]
0.00 0.12	CHEM,UV	2 hr	IE (68)	−1.1 −4.5 (p=0.016)[f] (range: +7 to −17)	Cough	22 male (22.3 ± 3.1 yr)	McDonnell et al.[16]
0.00 0.12	CHEM,UV	2 hr	IE (39)	−0.5 −3.4 (p=0.03) (range: +5 to −22)	None	23 male (8–11 yr)	McDonnell et al.[17]
0.00 0.14[e]	UV,UV	1 hr	CE (31)	−0.3 −4.2 (p<0.01)	None	46 male 13 female (12–15 yr)	Avol et al.[22,23]
0.00 0.15[e]	UV,UV	1 hr	CE (53)	+0.6 −5.3 (p<0.05)	Lower Respiratory Symptoms	42 male 8 female (26.4 ± 6.9 yr)	Avol et al.[18]

continued

Table 1. Continued

Ozone Concentration ppm	Measurement[a,b] Method	Exposure Duration	Activity[c] Level (V_E)	Group Mean % Change in FEV_1 (Range for O_3 Exp.)	Symptoms Reported	No., Sex and Age of Subjects	Reference
0.00 0.15	UV,UV	2 hr	IE (68)	+1.3 −0.7 (nd) (range: +5 to −10)	Cough	20 male (25.3 ± 4.1 yr)	Kulle et al.[21]
0.00 0.15	UV,UV	1 hr (mouthpiece)	CE (55)	+0.6 −4.5 (ns) (range: +3.5 to −31)		10 female (22.9 ± 2.5 yr)	Gibbons and Adams[28]
0.00 0.16	UV,UV	1 hr	CE (57)	+0.6 −6.1 (p<0.05)	Lower Respiratory Symptoms	42 male 8 female (26.4 ± 6.9 yr)	Avol et al.[18]
0.00 0.16[e]	UV,NBKI	1 hr	CE (38)	−0.1 −0.8 (ns)	No significant changes in symptom score	27 male 21 female (28 ± 8 yr)	Linn et al.[27] Avol et al.[26]
0.00 0.17[e]	UV,NBKI	1 hr	CE (42)	−0.4 −3.4 (p<0.006)	No significant changes in symptom score	45 male 15 female (30 ± 11 yr)	Linn et al.[27] Avol et al.[26]
0.00 0.17[e]	UV,NBKI	2 hr	IE (2×R)	+0.6 −2.1 (p<0.05)	Increased symptom scores	14 male 20 female (29 ± 8 yr)	Linn et al.[24]
0.00 0.18	CHEM,UV	2 hr	IE (65)	−1.1 −6.4 (p=0.008) (range: +2 to −22)	Cough	20 male (23.3 ± 2.8 yr)	McDonnell et al.[16]

continued

Table 1. Continued

Ozone Concentration ppm	Measurement[a,b] Method	Exposure Duration	Activity[c] Level (V_E)	Group Mean % Change in FEV_1 (Range for O_3 Exp.)	Symptoms Reported	No., Sex and Age of Subjects	Reference
0.0 0.2	UV,UV	2 hr	IE (68)	+1.3 −3.3 (nd) (range: +5 to −20)	Cough	20 male (25.3 ± 4.1 yr)	Kulle et al.[21]
0.0 0.2	UV,UV	2 hr	IE (30 for male, 18 for female subjects	+0.3 −3.1 (ns) (range: +6 to −17)		8 male 13 female (18–31 yr)	Gliner et al.[29]
0.00 0.21	UV,UV	1 hr	CE (80)	+1.9 −15 (p < 0.05)	Subjective symptoms which may limit performance	6 male 1 female (18–27 yr)	Folinsbee et al.[20]
0.00 0.24	CHEM,UV	2 hr	IE (65)	−1.1 −14.4 (p < 0.005) (range: −2 to −41)	Cough, pain on deep inspiration, shortness of breath	21 male (22.9 ± 2.9 yr)	McDonnell et al.[16]
0.00 0.24	UV,UV	1 hr	CE (60)	+0.6 −19 (p < 0.05)	Lower respiratory symptoms	42 male 8 female (26.4 ± 6.9 yr)	Avol et al.[18]

continued

Table 1. Continued

Ozone Concentration ppm	Measurement[a,b] Method	Exposure Duration	Activity[c] Level (V_E)	Group Mean % Change in FEV_1 (Range for O_3 Exp.)	Symptoms Reported	No., Sex and Age of Subjects	Reference
0.00 0.25	UV,UV	2 hr	IE (68)	+1.3 −6.7 (nd) (range: +5 to −35)	Cough, nose/ throat irritation	20 male (25.3 ± 4.1 yr)	Kulle et al.[21]

[a]Measurement method: CHEM = gas phase chemiluminescence; UV = ultraviolet photometry.
[b]Calibration method: NBKI = neutral buffered potassium iodide; UV = ultraviolet photometry.
[c]Minute ventilation reported in L/min or as a multiple of resting ventilation; IE = intermittent exercise; CE = continuous exercise.
[d]Pre to post difference (percent) in the group mean; statistical significance based on difference between O_3 and filtered air (0.0 ppm O_3) exposures: ns = not significant; nd = not determined.
[e]Measured in ambient air (mobile laboratory).
[f]"Suggested" significance based on Bonferroni inequality correction (p < 0.006).

Pulmonary function effects of chronic exposures to O_3 have been investigated in epidemiological and animal toxicological studies. Studies comparing communities have thus far been relatively unsuccessful, due to the lack of differences in pollutant levels, inadequate control of covariables, and insufficient individual exposure data. Chronic exposures for periods of days to months have resulted in increased end expiratory lung volume in adult rats at 0.25 ppm O_3,[39] increased pulmonary resistance in adult rats at 0.2 ppm O_3,[40] decreased lung compliance in adult monkeys at 0.5 ppm O_3,[41] decreased peak inspiratory flow in neonatal rats at 0.12 ppm,[39] and substantial decreases in lung function of adult monkeys at 0.64 ppm O_3, even after a 3 month post-exposure period.[42]

The weight of evidence clearly indicates that healthy, heavily exercising subjects experience statistically significant pulmonary function decrements during controlled exposures of ≥ 0.12 ppm O_3 for 2 hours or ambient exposures of ≥ 0.144 ppm O_3 for 1 hour. Level of exercise and individual responsiveness play a major role in determining the extent of pulmonary function impairment. Due to individual variability, potentially 5% to 20% of heavily exercising, health individuals may be sufficiently responsive to O_3 exposure to be considered at increased risk of pulmonary function and symptomatic effects (discussed below) at O_3 levels near the current O_3 NAAQS. Further review of controlled exposure data is underway in the Office of Air Quality Planning and Standards (U.S. EPA) to estimate the fraction of subjects responding excessively at O_3 concentrations in the range of 0.12 to 0.40 ppm.

Symptomatic Effects

Respiratory symptoms have been closely associated with pulmonary function changes in adults acutely exposed in controlled exposures to O_3 and in ambient air containing O_3 as the predominant pollutant. Although symptoms are less quantifiable than pulmonary function measurements, this association holds for both the time-course and magnitude of effects. Some symptoms, such as cough and chest pain, may interfere with maximal inspiration and expiration.

In controlled O_3 exposures, some heavily exercising (V_E = 65 L/min) adult subjects have experienced cough, shortness of breath, and pain on deep inspiration at 0.12 ppm O_3, although the group mean response was statistically significant for cough only.[16] Above 0.12 ppm O_3, respiratory and nonrespiratory symptoms which have been reported include throat dryness, chest tightness, substernal pain, cough, wheeze, pain on deep inspiration, shortness of breath, dyspnea, lassitude, malaise, headache, and nausea.[43-46] At 0.2 ppm O_3 and higher, controlled exposure studies of exercising subjects have reported positive ''symptom scores'' with elimination of symptoms for most subjects within 24 hours post-exposure.

Comparisons of symptoms reported in controlled O_3 exposure studies have been made with those reported in field studies. One study, which provided a direct comparison of symptoms in exercising (V_E = 57 L/min) adults caused by exposure to either purified air containing 0.16 ppm O_3 or oxidant-polluted ambient air which contained 0.15 ppm O_3, showed no significant differences, suggesting that increased symptoms associated with lung function impairment were caused only by O_3.[18] Several epidemiology studies have provided evidence of qualitative associations between ambient oxidant levels > 0.10 ppm and symptoms in children and young adults such as throat irritation, chest discomfort, cough, and headache.[49,50] Thus, it can be concluded that most symptoms reported in individuals exposed to O_3 in purified air are similar, but not identical, to those found for ambient air exposures.

An exception is eye irritation, a common symptom associated with exposure to photochemical oxidants, which has not been reported for controlled exposures to O_3 alone. This appears to hold even at O_3 concentrations much higher than would be found in the ambient air. It is widely accepted that other oxidants such as aldehydes and peroxyacetyl nitrate (PAN) are primarily responsible for eye irritation and are generally found in atmospheres containing higher ambient O_3 levels.[51-54]

Pulmonary function decrements have been reported in studies which do not report symptoms. Children (age 8–11) intermittently exercising (V_E = 39 L/min) for 2.5 hours at 0.12 ppm O_3 showed small, but statistically significant decreases in FEV_1 which persisted for 24 hours post-exposure but showed no changes in frequency or severity of cough compared to control.[17,47] Similarly, adolescents (age 12–15) continuously exercising (V_E = 31–33 L/min) during exposure to 0.144 ppm mean O_3 in ambient air showed no changes in symptoms despite statistically significant decrements in group mean FEV_1 (4%) which persisted at least one hour during resting post-exposure; the adolescents showed no changes in symptoms compared to the control.[22,23] Because symptoms can be viewed as an early warning of related lung function impairment by O_3 resulting in a "voluntary" limitation of activity, the lack of symptoms in children and adolescents during exposures which induce functional decrements may be of concern. This suggests that children may be at higher risk, since with no warning, they may not limit their activities during elevated O_3 episodes.

Symptoms when combined with objective measures of lung function are considered useful adjuncts in assessing health effects caused by O_3 and other photochemical oxidants. They have been shown to be closely associated with the time-course and magnitude of pulmonary function changes associated with O_3 exposures. Because symptoms caused by exposure to O_3 and other photochemical oxidants are associated with discomfort, interfere with normal activity, and provide subjective evidence of functional impairment, it has been recommended that they be considered health effects.

Exercise Performance

Early epidemiological evidence on high school students showed that the percentage of track team members failing to improve performance increased with increasing oxidant concentrations the hour before a race.[48] The authors concluded that the effects may have been related to increased R_{aw} or to associated discomfort which may have limited motivation to run at maximal levels. Controlled exposure studies of heavily exercising competitive runners have demonstrated decreased maximum V_E at 0.3 ppm O_3[55] and decreased FVC, FEF, and FEF at 0.20 ppm O_3.[19] At 0.21 ppm O_3, Folinsbee et al.[20] reported decreases in FVC, FEV, FEF, IC, and MVV at 75% maximum VO_2, as well as symptoms (laryngeal and tracheal irritation, soreness, and chest tightness on inspiration) in 7 distance cyclists exercising heavily (V_E = 81 L/min).

Too limited a database is available to draw any conclusive judgments regarding effects of O_3 on exercise performance. Although subjective statements by individuals engaged in sports indicate possible voluntary curtailing of activities during high-oxidant episodes, increased temperature and relative humidity may be involved in provoking the symptoms and lung function decrements observed in the studies above. Controlled studies of O_3 exposure have, however, demonstrated lung function impairment and subjective symptoms which cause individuals to reduce work load and performance. Because O_3 is implicated at least in part in reducing exercise performance during periods of high oxidants, this effect of O_3 has been viewed as a matter of public health concern.

Aggravation of Existing Respiratory Disease

Some epidemiological studies suggest an association between photochemical smog and aggravation of existing respiratory disease. No clear evidence is available, however, from controlled exposure or field studies to suggest that individuals with asthma, chronic bronchitis, or emphysema have greater lung function impairment caused by O_3 or other photochemical oxidants than healthy persons. Individuals with preexisting respiratory disease are considered to be especially "at risk" to O_3 exposure, due to their already compromised respiratory systems and concern that increased symptoms or pulmonary function decrements may interfere with normal function.

In controlled human exposure studies, statistically significant group mean decrements in pulmonary function were not reported for adult asthmatics exposed for 2 hours at rest[56], or with intermittent light exercise[57] to O_3 concentrations of 0.25 ppm or lower. Similarly, no statistically significant group mean changes in pulmonary function or symptoms were found in adolescent asthmatics exposed to 0.12 ppm O_3 for one hour at rest.[58] Subjects with chronic obstructive lung disease (COLD) performing light to moderate intermittent exercise show no statistically significant group mean decrements in pulmonary function during 1- and 2-hour exposures to ≤ 0.30 ppm O_3[59-61] and small group mean decrements in FEV_1 for 3-hour exposure of chronic bronchitics to 0.41 ppm O_3.[21] While these controlled exposure results suggest that individuals with preexisting respiratory disease may not be more sensitive to O_3 than healthy subjects, experimental design considerations in these studies suggest that the issues of sensitivity and aggravation of preexisting respiratory disease remain unresolved.

Epidemiological studies do not provide a clear concentration-response relationship between O_3 and aggravation of disease. For example, one study[62] conducted in Los Angeles reported increased daily asthma attack rates on cool days and days with high oxidants and particulates when median daily maximum hourly oxidant levels were ≤ 0.15 ppm; questionable exposure assessment, lack of control for medication, pollen, respiratory infections and other pollutants limits the use of this study for developing quantitative dose-response relationships.

Despite several exposure and variable control limitations, the Houston Area Oxidants Study (HAOS) concluded that for the study period in which the daily maximum hourly O_3 concentrations were < 0.21 ppm near the subjects' residence, (a) there was increased incidence of nasal and respiratory symptoms and increased frequency of medication use for asthmatics with increasing O_3 levels; (b) FEV_1 and FVC decreased with increasing O_3 and total oxidants; and (c) increased incidence of chest discomfort, eye irritation, and malaise occurred at high PAN concentrations.[63] In a subsequent related study, increased probability of an asthma attack was associated with the occurrence of a previous attack and with exposure to increased O_3 concentrations and temperature when maximum 1-hour averages for O_3 were between 0.001 and 0.127 ppm; however, other pollutants such as SO_2 and particulates may have been involved.[64-66]

In a series of studies conducted in a Tucson community, adults with asthma, allergies, or airway obstructive disease (AOD) were observed during an 11-month period in which 1-hour daily maximum O_3 concentrations were ≤ 0.12 ppm.[34-36] After adjusting for covariables, O_3 and TSP levels were significantly associated with peak expiratory flow rate in adults with AJD, and there was a significant interaction for O_3 and temperature with alterations in peak flow and symptoms in asthmatics. While these results suggest an effect of O_3 in individuals with preexisting respiratory disease, interpretation is difficult, due to the small sample size in relation to the number of covariates and the fact that individual exposure data were not available.

None of these epidemiology studies by themselves definitively demonstrates a relationship between O_3 and aggravation of preexisting respiratory disease. All of the studies report effects which may be related to inhalable particle exposure, and most have inadequate characterization of exposure. However, the group of studies as a whole support the contention that exposures to ambient levels of O_3 and other photochemical oxidants recently reported in many cities may increase the rate of asthma attacks, an effect which has been described as an adverse effect by the American Thoracic Society.[67]

Morphological Effects

Morphological effects of O_3 have been reported exclusively in animal toxicology studies. For this reason it is important to consider the differences in dosimetry and sensitivity between humans and laboratory animals. The following discussion provides an indication of some of the structural changes which might occur in human lungs as a result of repeated and long-term exposures of humans to O_3.

Despite the differences in lung structure between humans, dogs, monkeys, mice, rats, and guinea pigs, a characteristic lesion occurs at the junction of the conducting airways and the gaseous exchange tissues. In all species examined the effect is typically damage to the ciliated and Type 1 cells and hyperplasia of nonciliated bronchiolar and Type 2 cells; an increase in inflammatory cells is also observed. These effects were reported after 7 days, 8-hr/day exposures to 0.2 ppm O_3 in monkeys,[68,69] and in rats exposed for 7 days, 8- and 24-hr/day.[70] Similar effects with different exposures were reported in rats (0.26 ppm, 6 hr, endotracheal tube,[71] mice (0.5 ppm, 35 days[72]), and guinea pigs (0.5 ppm, 6 months[73]). Inadequate data limit quantitative comparisons between monkeys and rats, but a rough equivalency of responses has been observed under similar exposure conditions between species. Because all species tested show similar morphological responses to O_3 exposure, there is no reason to believe that humans exposed to O_3 would not respond similarly, although such effects may not necessarily occur at the same O_3 exposure concentrations or averaging times of exposure.

Changes in lung structure of monkeys and rats tend to decrease after extended exposure to O_3, although structural changes in the centriacinar region are reported after long-term exposures of rats,[74-77] monkeys,[78,79] and dogs.[80] While cell repair begins within 18 hours of exposure,[81-85] cell damage continues throughout long-term exposures, but at a slower rate.

Increases in lung collagen content are indicative of lung structure damage. Lung collagen content increases after short-term exposures to < 1.0 ppm O_3,[86,87] and continues to increase during long-term exposures.[88,89] Weanling and adult rats exposed for 6 and 13 weeks, respectively, and young monkeys for one year to < 1.0 ppm O_3 also show increased collagen content in the lungs.[90] In the latter study, examination of some of the exposed weanlings and controls at six weeks postexposure indicate a continued increase in lung collagen content, a result demonstrating that damage continues to occur during the postexposure period.

The centriacinar inflammatory process also continues during longterm O_3 exposures and appears to be related to remodeling of the centriacinar airways[74,75,79] and to increased lung collagen which appears mainly in the centriacinar regions.[90] In addition there is morphometric,[79] morphologic,[80] and functional[40,42] evidence of distal airway narrowing.

Implications for humans of these lung structure changes observed in animals are unclear at present. While the distal airway narrowing and lesions at the junction of the conducting airways and gaseous exchange region are similar to the changes which have been found in lungs of cigarette smokers,[91-94] there is no evidence of emphysema in the lungs of animals exposed to only O_3. Many of the lung structural changes that have been reported for long-term exposures to < 1.0 ppm O_3 in several different species are considered adverse and relevant to standard setting if demonstrated in humans at ambient exposure concentrations of O_3. Although the lowest O_3 concentrations and shortest averaging times which could produce structural changes in human lungs is uncertain at this time, there appears to be a need to protect the public from O_3 exposures which may induce such health effects.

Altered Host Defense Systems

Respiratory systems of mammals are protected from bacterial and viral infections by the closely interrelated particle removal (both mucociliary and phagocytic) and immunological defense systems. Numerous factors such as poor nutrition, preexisting disease, and environmental stress may influence or alter these host defense systems of individuals sufficiently to permit development of respiratory infections. Animal studies indicate that exposure to O_3 is one of those factors.

Both in vivo (live animal) and in vitro (isolated cell) studies demonstrate that O_3 can affect the ability of the clearance and immune systems to defend against infection. Increased susceptibility to bacterial infection is reported in mice at 0.08 to 0.10 ppm O_3 for a single 3-hour exposure,[95-97] and at 0.10 ppm O_3 for subchronic exposures.[98] Several related alterations of the pulmonary defenses caused by short-term and subchronic exposures to O_3 include: (a) impaired ability to inactivate bacteria in rabbits and mice;[99-103] impaired performance of mucociliary clearance mechanisms;[104-107] (b) immuno-suppression;[108-110] (c) significantly reduced number of pulmonary defense cells in rabbits;[99,111] and (d) impaired macrophage phagocytic activity, less macrophage mobility, more fragility and membrane alterations, and reduced lysosomal enzymatic activity.[112-120] Some of these effects are shown to occur in a variety of species, including mice, rats, rabbits, guinea pigs, dogs, sheep, and monkeys.

Other studies indicate similar effects for short-term and subchronic exposures of mice to O_3 combined with pollutants such as SO_2, NO_2, H_2SO_4, and particles.[121-124] Similar to human pulmonary function response to O_3, activity levels of mice exposed to O_3 are shown to play a role in determining the lowest effective concentration which alters the immune defenses.[125]

Although this large body of evidence clearly demonstrates that short-term and subchronic exposures to O_3 can impair the immune defense systems of animals, technical and ethical considerations have limited similar research on human subjects. Thus inferences have been drawn and models developed to assess the relevance of animal data to humans. For example, animal endpoints such as increased mortality could be appropriately compared to increased morbidity in humans, such as the increased prevalence of respiratory illness in the community.

Based on current understanding of physiology, metabolism, and immune defenses, it is generally accepted that the basic mechanisms of action associated with defense against infectious agents are similar in humans and animals. Similarities between human and rodent antibacterial systems have been discussed in detail,[126] but differences in lung structure and biological response will inevitably cause differences in dosimetry, sensitivity, and endpoints of response. In addition, other factors such as preexisting disease, nutrition, presence of other pollutants, and environmental stresses (e.g., nitrogen dioxide, sulfur dioxide, particulate matter, high temperature or humidity) can influence the effect of exposure to O_3 and infectious agents. Despite the differences in related factors which make precise estimation of human response from animal data difficult, it is reasonable to hypothesize that humans exposed to O_3 could experience impairment of host defenses.

Extrapulmonary Effects

Extrapulmonary effects which are demonstrated in humans or laboratory animals following exposure to O_3 include alterations in red blood cell morphology and enzyme activity, cytogenetic effects in circulating lymphocytes, and subjective limitations in vigilance tasks. Additionally, animal toxicology studies provide limited evidence for cardiovascular, reproductive, teratological, endocrine system, and liver metabolism effects. This wide variety of extrapulmonary effects is probably caused by oxidative reaction products of O_3. Due to the high reactivity of O_3 with biological tissue, mathematical models predict that only a small fraction of O_3 actually reaches the circulatory system.[127]

Of the extrapulmonary effects reported, cytogenetic and mutational effects are probably the most controversial. Statistically significant increases in frequency of sister chromatid exchanges (chromosomal alterations) have been caused by in vitro exposures to 0.25 ppm for 1 hour,[128] suggesting a mutagenic response. However, in vitro responses are not extrapolatable to in vivo responses, because homeostatic mechanisms are not represented. Therefore, isolated in vitro exposure studies cannot be used to provide accurate estimations of risk. In vivo animal studies have shown significant increases in the number of chromosomal and chromatid aberrations following 4 and 5 hour exposures to 0.2 and 0.43 ppm O_3, respectively.[129-131] However, animal studies[132] and controlled human exposures to O_3 levels as high as 0.5 ppm have shown no significant cytogenetic effects attributable to O_3,[133-135] and epidemiology studies provide no evidence of chromosomal changes induced by ambient O_3.[136] Thus, while the animal studies are suggestive of possible cytogenetic effects from O_3, human studies have not demonstrated such effects for realistic human exposures to O_3.

Limited hematological and serum chemistry effects data indicate that O_3 can interfere with biochemical mechanisms in human blood erythrocytes and sera, but the physiological significance of these studies is unclear. While the behavior of rats is significantly affected during exposure to concentrations as low as 0.12 ppm O_3 for 6 hours, it is unknown whether lung irritation, odor, or a direct effect on the central nervous system (CNS) causes change in rodent behavior at lower O_3 concentrations. Other O_3-induced effects such as pentobarbital-induced sleep time alterations and hormone level changes have been reported and hypotheses made suggesting that there may be a cause-effect relationship between

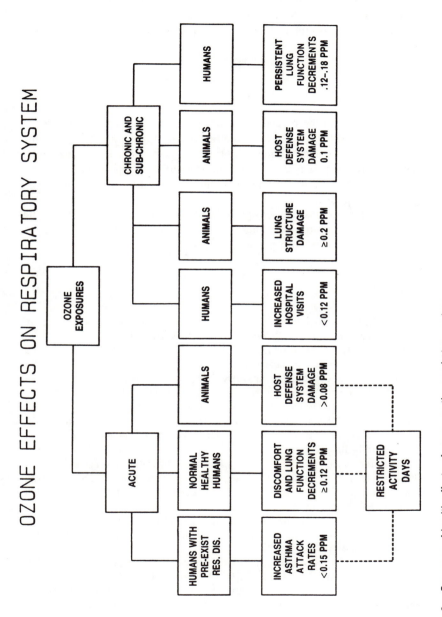

Figure 3.   Summary of health effects of ozone on the respiratory system.

ozone exposure and such changes. With regard to other extrapulmonary effects, exact mechanisms remain to be elucidated, and physiological significance for humans remains uncertain. Nonetheless, the body of this evidence suggests that O_3 can cause effects distant from the lungs in animals, and hence possibly in humans.

SUMMARY

As can be seen from the preceding review of O_3 health effects literature, O_3 has many and varied health effects on the respiratory system and possibly other systems of the body. Because effects of O_3 on the respiratory system have been documented more completely than extrapulmonary effects, respiratory effects tend to be the focus of regulatory decision-making. Respiratory effects of ozone are summarized in Figure 3.

A new emerging database as exemplified by the work of Lioy et al.[38] indicates a need for consideration of other longer (e.g., 7 or 8 hour) averaging times for an additional or alternative O_3 primary standard.

REFERENCES

1. Padgett, J., and H. Richmond. "The Process of Establishing National Ambient Air Quality Standards," *J. Air Pollut. Control.* 33:13–16 (1983).
2. Jordan, B.C., H. M. Richmond, and T. McCurdy. "The Use of Scientific Information in Setting Ambient Air Standards," *Environ. Health Perspec.* 52:233–240 (1983).
3. Lippman, M. "Role of Science Advisory Groups in Establishing Standards for Ambient Air Pollutants," *Aerosol Sci. and Tech.* 6:93–114 (1987).
4. U.S. Environmental Protection Agency. "Air Quality Criteria for Ozone and Other Photochemical Oxidants," Research Triangle Park, NC: U.S. Environmental Protection Agency, Environmental Criteria and Assessment Office; EPA-600/8-841020aF–020eF (1986).
5. U.S. Environmental Protection Agency. "Review of the National Ambient Air Quality Standards for Ozone: Preliminary Assessment of Scientific and Technical Information," Durham, NC: U.S. Environmental Protection Agency, Office of Air Quality Planning and Standards (1989).
6. Silverman, F., L. J. Folinsbee, J. Barnard, and R. J. Shephard. "Pulmonary Function Changes in Ozone—Interaction of Concentration and Ventilation," *J. Appl. Physiol.* 41:859–864 (1976).
7. Folinsbee, L. J., F. Silverman, and R. J. Shephard. "Exercise Responses Following Ozone Exposure," *J. Appl. Physiol.* 38:996–1001 (1975).
8. Bates, D. V., G. M. Bell, C. D. Burnham, M. Hazucha, J. Mantha, L. D. Pengelly, and F. Silverman. "Short-Term Effects of Ozone on the Lung," *J. Appl. Physiol.* 32:176–181 (1972).
9. Golden, J. A., J. A. Nadel, and H. A. Boushey. "Bronchial Hyperirritability in Healthy Subjects After Exposure to Ozone," *Am. Rev. Respir. Dis.* 118:287–294 (1978).

10. Folinsbee, L. J., B. L. Drinkwater, J. F. Bedi, and S. M. Horvath. "The Influence of Exercise on the Pulmonary Changes Due to Exposure to Low Concentrations of Ozone," in L. J. Folinsbee, J. A. Wagner, J. F. Borgia, B. L. Drinkwater, J. A. Gliner, and J. F. Bedi, eds. *Environmental Stress: Individual Human Adaptations* (New York, NY: Academic Press, 1978), pp. 125–145.

11. Horvath, S. M., J. A. Gliner, and J. A. Matsen-Twisdale. "Pulmonary Function and Maximum Exercise Responses Following Acute Ozone Exposure," *Aviat. Space Environ. Med.* 50:901–905 (1979).

12. Shephard, R. J., B. Urch, F. Silverman, and P. N. Corey. (1983) "Interaction of Ozone and Cigarette Smoke Exposure," *Environ. Res.* 31:125–137 (1983).

13. Koenig, G., H. Rommelt, H. Kienele, K. Dirnagl, H. Polke, and G. Fruhmann. "Changes in the Bronchial Reactivity of Humans Caused by the Influence of Ozone," *Arbeitsmed. Sozialmed. Praeventivmed* 151:261–263 (1980).

14. Niinimaa, V., P. Cole, S. Mintz, and R. J. Shephard. "The Switching Point from Nasal to Oronasal Breathing," *Respir. Physiol.* 42:61–71 (1980).

15. Niinimaa, V., P. Cole, S. Mintz, and R. J. Shephard. "Oronasal Distribution of Respiratory Airflow," *Respir. Physiol.* 43:69–75 (1981).

16. McDonnell, W. F., D. H. Horstmann, M. J. Hazucha, E. Seal, Jr., E. D. Haak, S. Salaam, and D. E. House. "Pulmonary Effects of Ozone Exposure During Exercise: Dose-Response Characteristics," *J. Appl. Physiol.: Respir. Environ. Exercise Physiol.* 54:1345–1352 (1983).

17. McDonnell, W. F., III, R. S. Champan, M. W. Leigh, G. L. Strope, and A. M. Collier. "Respiratory Responses of Vigorously Exercising Children to 0.12 ppm Ozone Exposure," *Am. Rev. Respir. Dis.* 132:875–879 (1985b).

18. Avol, E. L., W. S. Linn, T. G. Venet, D. A. Shamoo, and J. D. Hackney. "Comparative Respiratory Effects of Ozone and Ambient Oxidant Pollution Exposure During Heavy Exercise," *J. Air Pollut. Control Assoc.* 34: 804–809 (1984).

19. Adams, W. C., and E. S. Schelegle. "Ozone and High Ventilation Effects on Pulmonary Function and Endurance Performance," *J. Appl. Physiol.: Respir. Environ. Exercise Physiol.* 55:805–812 (1983).

20. Folinsbee, L. J., J. F. Bedi, and S. M. Horvath. "Pulmonary Function Changes in Trained Athletes Following 1-Hour Continuous Heavy Exercise While Breathing 0.21 ppm Ozone," *J. Appl. Physiol.: Respir. Environ. Exercise Physiol.* 57: 984–988 (1984).

21. Kulle, T. J., L. R. Sauder, J. R. Hebel, and M. D. Chatham. "Ozone Response Relationships in Healthy Nonsmokers," *Am. Rev. Respir. Dis.* 132:36–41 (1985).

22. Avol, E. L., W. S. Linn, D. A. Shamoo, L. M. Valencia, U. T. Anzar, and J. D. Hackney. "Respiratory Effects of Photochemical Oxidant Air Pollution in Exercising Adolescents," *Am. Rev. Respir. Dis.* (1985a).

23. Avol, E. L., W. S. Linn, D. A. Shamoo, L. M. Valencia, U. T. Anzar, and J. D. Hackney. "Short-Term Health Effects of Ambient Air Pollution in Adolescents," in *Proceedings of an International Specialty Conference on the Evaluation of the Scientific Basis for an Ozone/Oxidant Standard;* November 1984; Houston, TX. (Pittsburgh, PA: Air Pollution Control Association, 1985b), pp. 329–336. (APCA international specialty conference transactions: v. 4).

24. Linn, W. S., M. P. Jones, E. A. Bachmayer, C. E. Spier, S. F. Mazur, E. L. Avol, and J. D. Hackney. "Short-Term Respiratory Effects of Polluted Air: a Laboratory Study of Volunteers in a High-Oxidant Community," *Am. Rev. Respir. Dis.* 121:243–252 (1980).

25. Linn, W. S., D. A. Shamoo, T. G. Venet, C. E. Spier, L. M. Valencia, U. T. Anzar, and J. D. Hackney. "Response to Ozone in Volunteers with Chronic Obstructive Pulmonary Disease," *Arch. Environ. Health* 38:278–283 (1983b).

26. Avol, E.L., W. S. Linn, D. A. Shamoo, T. G. Venet, and J. D. Hackney. "Acute Respiratory Effects of Los Angeles Smog in Continuously Exercising Adults," *J. Air Pollut. Control Assoc.* 33:1055–1060 (1983).

27. Linn, W. S., E. L. Avol, and J. D. Hackney. "Effects of Ambient Oxidant Pollutants on Humans: A Movable Environmental Chamber Study," in S. D. Lee, M. G. Mustafa, and M. A. Mehlman, eds. *International Symposium on the Biomedical Effects of Ozone and Related Photochemical Oxidants,* March 1982, Pinehurst, NC. (Princeton, NJ: Princeton Scientific Publishers, Inc., 1983a), pp. 125–137. (Advances in Modern Toxicology: v. 5).

28. Gibbons, S. I., and W. C. Adams. "Combined Effects of Ozone Exposure and Ambient Heat on Exercising Females," *J. Appl. Physiol.* 57:450–456 (1984).

29. Gliner, J. A., S. M. Horvath, and L. J. Folinsbee. "Pre-exposure to Low Ozone Concentrations Does Not Diminish the Pulmonary Function Response on Exposure to Higher Ozone Concentration," *Am. Rev. Respr. Dis.* 127:51–55 (1983).

30. Hayes, S.R., et al. "An Analysis of Symptom and Lung Function Data from Several Human Controlled Ozone Exposure Studies," Systems Applications, Inc. for U.S. Environmental Protection Agency, Research Triangle Park, NC; March 25, 1987.

31. Kagawa, J., and T. Toyama. "Photochemical Air Pollution: Its Effects on Respiratory Function of Elementary School Children," *Arch. Environ. Health* 30:117–122 (1975).

32. Kagawa, J., T. Toyama, and M. Nakaza. "Pulmonary Function Test in Children Exposed to Air Pollution," in A. J. Finkel, and W. C. Duel, eds. *Clinical Implications of Air Pollution Research: Proceedings of the* 1974 *Air Pollution Medical Research Conference,* December 1974, San Francisco, CA. (Acton, MA: Publishing Sciences Group, Inc., 1976), pp. 305–320.

33. Lippman, M., P. J. Lioy, G. Leikauf, K. B. Green, D. Baxter, M. Morandi, and B. S. Pasternack. "Effects of Ozone on the Pulmonary Function of Children," in S. D. Lee, M. G. Mustafa, and M. A. Mehlman, eds. *International Symposium on the Biomedical Effects of Ozone and Related Photochemical Oxidants,* March 1982, Pinehurst, NC. (Princeton, NJ: Princeton Scientific Publishers, Inc., 1983), pp. 423–446. (Advances in Modern Toxicology: v. 5).

34. Lebowitz, M. D., M. K. O'Rourke, R. Dodge, C. J. Holberg, G. Corman, R. W. Hoshaw, J. L. Pinnas, R. A. Barbee, and M. R. Sneller. "The Adverse Health Effects of Biological Aerosols, Other Aerosols, and Indoor Microclimate on Asthmatics and Nonasthmatics," *Environ. Int.* 8:375–380 (1982).

35. Lebowitz, M. D., C. J. Holberg, and R. R. Dodge. "Respiratory Effects on Populations from Low Level Exposures to Ozone," Presented at the 34th annual meeting of the Air Pollution Control Association, June, Atlanta, GA. (Pittsburgh, PA: Air Pollution Control Association; paper no. 83-12.5, 1983).

36. Lebowitz, M. D. "The Effects of Environmental Tobacco Smoke Exposure and Gas Stoves on Daily Peak Flow Rates in Asthmatic and Non-Asthmatic Families," *Eur. J. Respir. Dis.* 65 (suppl. 133): 90–97 (1984).

37. Bock, N., M. Lippman, P. Lioy, A. Munoz, and F. Speizer. "The Effects of Ozone on the Pulmonary Function of Children," in *Proceedings of an International Specialty Conference on the Evaluation of the Scientific Basis for an Ozone/Oxidant Standard,*

November 1984, Houston, TX. (Pittsburgh, PA: Air Pollution Control Association, 1985) pp. 297–308. (APCA International Specialty Conference Transactions: v. 4).

38. Lioy, P. J., T. A. Vollmuth, and M. Lippman. "Persistence of Peak Flow Decrement in Children Following Ozone Exposures Exceeding the National Ambient Air Quality Standard," *J. Air Pollut. Control Assoc.* 35:1068–1071 (1985).

39. Raub, J. A., F. J. Miller, and J. A. Graham. "Effects of Low-Level Ozone Exposure on Pulmonary Function in Adult and Neonatal Rats," in S. D. Lee, M. G. Mustafa, and M. A. Mehlman, eds. *International Symposium on the Biomedical Effects of Ozone and Related Photochemical Oxidants,* March 1982, Pinehurst, NC. (Princeton, NJ: Princeton Scientific Publishers, Inc., 1983), pp. 363–367. (Advances in Modern Environmental Toxicology: v. 5).

40. Costa, D. L., R. S. Kutzman, J. R. Lehmann, E. A. Popenoe, and R. T. Drew. "A Subchronic Multi-Dose Ozone Study in Rats," in S. D. Lee, M. G. Mustafa, and M. A. Mehlman, eds. *International Symposium on the Biomedical Effects of Ozone and Related Photochemical Oxidants,* March 1982, Pinehurst, NC. (Princeton, NJ: Princeton Scientific Publishers, Inc., 1983), pp. 369–393. (Advances in Modern Environmental Toxicology: v. 5.).

41. Eustis, S. L., L. W. Schwartz, P. C. Kosch, and D. L. Dungworth. "Chronic Bronchiolitis in Nonhuman Primates After Prolonged Ozone Exposure," *Am. J. Pathol.* 105:121–137 (1981).

42. Wegner, C. D. "Characterization of Dynamic Respiratory Mechanics by Measuring Pulmonary and Respiratory System Impedances in Adult Bonnet Monkeys (*Macaca radiata*): Including the Effects of Long-Term Exposure to Low-Level Ozone," [PhD dissertation]. (Davis, CA: University of California, 1982). Available from: University Microfilms, Ann Arbor, MI; publication no. 82-27900.

43. DeLucia, A. J., and W. C. Adams. "Effects of O_3 Inhalation During Exercise on Pulmonary Function and Blood Biochemistry," *J. Appl. Physiol.: Respir. Environ. Exercise Physiol.* 43:75–81 (1977).

44. Kagawa, J., and K. Tsuru. "Effects of Ozone and Smoking Alone and in Combination on Bronchial Reactivity to Inhaled Acetylcholine," *Nippon Kyobu Shikkan Gakkai Zasshi* 17:703–709 (1979a).

45. Kagawa, J., and K. Tsuru. "Respiratory Effects of 2-Hour Exposure to Ozone and Nitrogen Dioxide Alone and in Combination in Normal Subjects Performing Intermittent Exercise," *Nippon Kyobu Shikkan Gakkai Zasshi* 17:765–774 (1979b).

46. Kagawa, J., and K. Tsuru. "Respiratory Effect of 2-Hour Exposure with Intermittent Exercise to Ozone and Sulfur Dioxide Alone and in Combination in Normal Subjects," *Nippon Eiseigaku Zasshi* 34:690–696 (1979c).

47. Hammer, D. I., V. Hasselblad, B. Portnoy, and P. F. Wehrle. "Los Angeles Student Nurse Study: Daily Symptom Reporting and Photochemical Oxidants," *Arch. Environ. Health* 28:255–260 (1974).

48. Makino, K., and I. Mizoguchi. "Symptoms Caused by Photochemical Smog," *Nippon Koshu Eisei Zasshi* 22:421–430 (1975).

49. Altshuller, A. P. "Eye Irritation as an Effect of Photochemical Air Pollution," *J. Air Pollut. Control Assoc.* 27:1125–1126 (1977).

50. National Research Council. "Toxicology," in *Ozone and Other Photochemical Oxidants* (Washington, DC: National Academy of Sciences, Committee on Medical and Biologic Effects of Environmental Pollutants, 1977), pp. 323–387.

51. U.S. Environmental Protection Agency. "Air Quality Criteria for Ozone and Photochemical Oxidants," Research Triangle Park, NC: U.S. Environmental Protection Agency, Environmental Criteria and Assessment Office; EPA-600/8-78-004. Available from: NTIS, Springfield, VA; PB80-124753 (1978).

52. Okawada, N., I. Mizoguchi, and T. Ishiguro. "Effects of Photochemical Air Pollution on the Human Eye—Concerning Eye Irritation, Tear Lysome and Tear pH," *Nagoya J. Med. Sci.* 41:9–20 (1979).

53. McDonnell, W. F., III, D. H. Horstman, S. Abdul-Salaam, and D. E. House. "Reproducibility of Individual Responses to Ozone Exposure," *Am. Rev. Respir. Dis.* 131:36–40 (1985a).

54. Wayne, W. S., P. F. Wehrle, and R. E. Carroll. "Oxidant Air Pollution and Athletic Performance," *J. Am. Med. Assoc.* 199: 901–904 (1967).

55. Savin, W., and W. Adams. "Effects of Ozone Inhalation on Work Performance and VO_2 Max," *J. Appl. Physiol.* 46:309–314 (1979).

56. Silverman, F. "Asthma and Respiratory Irritants (Ozone)," *EHP Environ. Health Perspect.* 29:131–136 (1979).

57. Linn, W. S., R. D. Buckley, C. E. Spier, R. L. Blessey, M. P. Jones, D. A. Fischer, and J. D. Hackney. "Health Effects of Ozone Exposure in Asthmatics," *Am. Rev. Respir. Dis.* 117:835–843 (1978).

58. Koenig, J.Q., D. S. Covert, M. S. Morgan, N. Horike, S. G. Marshall, and W. E. Pierson. "Acute Effects of 0.12 ppm Ozone or 0.12 ppm Nitrogen Dioxide on Pulmonary Function in Healthy and Asthmatic Adolescents," *Am. Rev. Respir. Dis.* 132:648–651 (1985).

59. Linn, W. S., D. A. Medway, U. T. Anzar, L. M. Valencia, C. E. Spier, F. S-O. Tsao, D. A. Fischer, and J. D. Hackney. "Persistence of Adaptation to Ozone in Volunteers Exposed Repeatedly Over Six Weeks," *Am. Rev. Respir. Dis.* 125:491–495 (1982).

60. Kehrl, H. R., M. J. Hazucha, J. Solic, J.and P. A. Bromberg. "The Acute Effects of 0.2 and 0.3 ppm Ozone in Persons with Chronic Obstructive Lung Disease (COLD)," in S. D. Lee, M. G. Mustafa, and M. A. Mehlman, eds. *International Symposium on the Biomedical Effects of Ozone and Related Photochemical Oxidants,* March 1982, Pinehurst, NC. (Princeton, NJ: Princeton Scientific Publishers, Inc., 1983), pp. 213–225. (Advances in Modern Environmental Toxicology: v. 5).

61. Kehrl, H. R., M. J. Hazucha, J. J. Solic, and P. A. Bromberg. "Responses of Subjects with Chronic Pulmonary Disease After Exposures to 0.3 ppm Ozone," *Am. Rev. Respir. Dis.* 131:719–724 (1985).

62. Whittemore, A.S., and E. L. Korn. "Asthma and Air Pollution in the Los Angeles Area," *Am. J. Public Health* 70:687–696 (1980).

63. Javitz, H. S., R. Kransnow, C. Thompson, K. M. Patton, D. E. Berthiaume, and A. Palmer. "Ambient Oxidant Concentrations in Houston and Acute Health Symptoms in Subjects with Chronic Obstructive Pulmonary Disease: A Reanalysis of the HAOS Health Study," in S. D. Lee, M. G. Mustafa, and M. A. Mehlman, eds. *International Symposium on the Biomedical Effects of Ozone and Related Photochemical Oxidants,* March 1982, Pinehurst, NC. (Princeton, NJ: Princeton Scientific Publishers, Inc., 1983), pp. 227–256. (Advances in Modern Toxicology: v. 5).

64. Stock, T. H., A. H. Holguin, B. J. Selwyn, B. P. Hsi, C. F. Contant, P. A. Buffler, and D. J. Kotchmar. "Exposure Estimates for the Houston Area Asthma and

Runners Studies,'' in S. D. Lee, M. G. Mustafa, and M. A. Mehlman, eds. *Inter-national Symposium on the Biomedical Effects of Ozone and Related Photochemical Oxidants*, March 1982, Pinehurst, NC (Princeton, NJ: Princeton Scientific Publishers, Inc., 1983) (Advances in Modern Environmental Toxicology: v. 5).

65. Holguin, A. H., P. A. Buffler, C. R. Contant, Jr., T. H. Stock, D. J. Kotchmar, B. P. Hsi, D. E. Jenkins, B. M. Gehan, L. M. Noel, and M. Mei. ''The Effects of Ozone on Asthmatics in the Houston Area,'' in *Proceedings of an International Specialty Conference on the Evaluation of the Scientific Basis for an Ozone/Oxidant Standard*, November 1984, Houston, TX. (Pittsburgh, PA: Air Pollution Control Association, 1985), pp. 262–280. (APCA International Specialty Conference Trans-actions: v. 4).

66. Contant, C. F., Jr., B. M. Gehan, T. H. Stock, A. H. Holguin, and P. A. Buffler. ''Estimation of Individual Ozone Exposures using Microenvironment Measures,'' in *Proceedings of an International Specialty Conference on the Evaluation of the Scientific Basis for an Ozone/Oxidant Standard*, November 1984, Houston, TX. (Pittsburgh, PA: Air Pollution Control Association, 1985), pp. 250–261. (APCA International Specialty Conference Transactions: v. 4).

67. American Thoracic Society. ''Guidelines As To What Constitutes an Adverse Respira-tory Health Effect, With Special Reference to Epidemiologic Studies of Air Pollu-tion,'' *Am. Rev. Respir. Dis.* 131:666–668 (1985).

68. Dungworth, D. L., W. L. Castleman, C. K. Chow, P. W. Mellick, M. G. Mustafa, B. Tarkington, and W. S. Tyler. ''Effect of Ambient Levels of Ozone on Monkeys,'' *Fed. Proc. Fed. Am. Soc. Exp. Biol.* 34:1670–1674 (1975).

69. Castleman, W. L., W. S. Tyler, and D. L. Dungworth. ''Lesions in Respiratory Bronchioles and Conducting Airways of Monkeys Exposed to Ambient Levels of Ozone,'' *Exp. Mol. Pathol.* 26:384–400 (1977).

70. Schwartz, L.W., D. L. Dungworth, M. G. Mustafa, B. K. Tarkington, and W. S. Tyler. ''Pulmonary Responses of Rats to Ambient Levels of Ozone: Effects of 7-Day Intermittent or Continuous Exposure,'' *Lab. Invest.* 34:565–578 (1976).

71. Boatman, E. S., S. Sato, and R. Frank. ''Acute Effects of Ozone on Cat Lungs II. Structural,'' *Am. Rev. Respir. Dis.* 110:157–169 (1974).

72. Zitnik, L. A., L. W. Schwartz, N. K. McQuillen, Y. C. Zee, and J. W. Osebold. ''Pulmonary Changes Induced by Low-Level Ozone: Morphological Observations,'' *J. Environ. Pathol. Toxicol.* 1:365–376 (1978).

73. Cavender, F. L., B. Singh, and B. Y. Cockrell. ''Effects in Rats and Guinea Pigs of Six-Month Exposures to Sulfuric Acid Mist, Ozone, and Their Combination,'' *J. Toxicol. Environ. Health* 4: 845–852 (1978).

74. Boorman, G. A., L. W. Schwartz, and D. L. Dungworth. ''Pulmonary Effects of Prolonged Ozone Insult in Rats: Morphometric Evaluation of the Central Acinus,'' *Lab. Invest.* 43:108–115 (1980).

75. Moore, P. F., and L. W. Schwartz. ''Morphological Effects of Prolonged Exposure to Ozone and Sulfuric Acid Aerosol on the Rat Lung,'' *Exp. Mol. Pathol.* 35:108–123 (1981).

76. Barry, B. E., F. J. Miller, and J. D. Crapo. ''Alveolar Epithelial Injury Caused by Inhalation of 0.25 ppm of Ozone,'' in S. D. Lee, M. G. Mustafa, M. A. Mehl-man, eds. *International Symposium on the Biomedical Effects of Ozone and Related Photochemical Oxidants*, March 1982; Pinehurst, NC. (Princeton, NJ: Princeton

Scientific Publishers, Inc., 1983), pp. 299–309. (Advances in Modern Environmental Toxicology: v. 5).

77. Crapo, J. D., B. E. Barry, L.-Y. Chang, and R. R. Mercer. Alterations in Lung Structure Caused by Inhalation of Oxidants,'' *J. Toxicol. Environ. Health* 13:301–321 (1984).

78. Fujinaka, L. E. ''Respiratory Bronchiolitis Following Long-Term Ozone Exposure in Bonnet Monkeys: A Morphometric Study'' [master's thesis]. (Davis, CA: University of California, 1984).

79. Fujinaka, L. E., D. M. Hyde, C. G. Plopper, W. S. Tyler, D. L. Dungworth, and L. O. Lollini. ''Respiratory Bronchiolitis Following Long-Term Ozone Exposure in Bonnet Monkeys: A Morphometric Study,'' *Exp. Lung Res.* 8:167–190 (1985).

80. Freeman, G., R. J. Stephens, D. L. Coffin, and J. R. Stara. ''Changes in Dogs' Lungs After Long-Term Exposure to Ozone: Light and Electron Microscopy,'' *Arch. Environ. Health* 26:209–216 (1973).

81. Castleman, W.L., D. L. Dungworth, L. W. Schwartz, and W. S. Tyler. ''Acute Respiratory Broncholitis: An Ultrastructural and Autoradiographic Study of Epithelial Cell Injury and Renewal in Rhesus Monkeys Exposed to Ozone,'' *Am. J. Pathol.* 98:811–840 (1980).

82. Evans, M. J., L. V. Johnson, R. J. Stephens, and G. Freeman. ''Renewal of the Terminal Bronchiolar Epithelium in the Rat Following Exposure to NO_2 or O_3,'' *Lab. Invest.* 35:246–257 (1976a).

83. Evans, M. J., L. V. Johnson, R. J. Stephens, and G. Freeman. ''Cell Renewal in the Lungs of Rats Exposed to Low Levels of Ozone,'' *Exp. Mol. Pathol.* 24:70–83 (1976b).

84. Evans, M. J., R. J. Stephens, and G. Freeman. ''Renewal of Pulmonary Epithelium Following Oxidant Injury,'' in A. Bouhuys, ed. *Lung Cells in Disease: Proceedings of a Brook Lodge Conference,* April, 1976, Augusta, MI. (New York, NY: Elsevier/North-Holland Biomedical Press, 1976c), pp. 165–178.

85. Lum, H., L. W. Schwartz, D. L. Dungworth, and W. S. Tyler, ''A Comparative Study of Cell Renewal After Exposure to Ozone or Oxygen: Response of Terminal Bronchiolar Epithelium in the Rat,'' *Am. Rev. Respir. Dis.* 118:335–345 (1978).

86. Last, J. A., D. B. Greenberg, and W. L. Castleman. ''Ozone-Induced Alterations in Collagen Metabolism of Rat Lungs,'' *Toxicol. Appl. Pharmacol.* 51:247–258 (1979).

87. Last, J. A., T. W. Hesterberg, K. M. Reiser, C. E. Cross, T. C. Amis, C. Gunn, E. P. Steffey, J. Grandy, and R. Henrickson. ''Ozone-Induced Alterations in Collagen Metabolism of Monkey Lungs: Use of Biopsy-Obtained Lung Tissue,'' *Toxicol. Appl. Pharmacol.* 60:579–585 (1981).

88. Last, J. A., and D. B. Greenberg. ''Ozone-Induced Alterations in Collagen Metabolism of Rat Lungs. II. Long-Term Exposure,'' *Toxicol. Appl. Pharmacol.* 55:108–114 (1980).

89. Last, J. A., D. M. Hyde, and D. P. Y. Chang. ''A Mechanism of Synergistic Lung Damage by Ozone and a Respirable Aerosol,'' *Exp. Lung Res.* 7:223–235 (1984a).

90. Last, J. A., K. M. Reiser, W. S. Tyler, and R. B. Rucker. ''Long-Term Consequences of Exposure to Ozone: I. Lung Collagen Content,'' *Toxicol. Appl. Pharmacol.* 72:111–118 (1984b).

91. Niewoehner, D. E., J. Kleinerman, and D. B. Rice. ''Pathologic Changes in the Peripheral Airways of Young Cigarette Smokers,'' *N. Engl. J. Med.* 291:755–758 (1974).

92. Cosio, M. G., K. A. Hale, and D. E. Niewoehner. "Morphologic and Morphometric Effects of Prolonged Cigarette Smoking on the Small Airways," *Am. Rev. Respir. Dis.* 122:265–271 (1980).

93. Hale, K. A., D. E. Niewoehner, and M. G. Cosio. "Morphologic Changes in Muscular Pulmonary Arteries: Relationship to Cigarette Smoking, Airway Disease, and Emphysema," *Am. Rev. Respir. Dis.* 122:273–280 (1980).

94. Wright, J. L., L. M. Lawson, P. D. Pare, B. J. Wiggs, S. Kennedy, and J. C. Hogg. "Morphology of Peripheral Airways in Current Smokers and Exsmokers," *Am. Rev. Respir. Dis.* 127:474–477 (1983).

95. Coffin, D. L., E. J. Blommer, D. E. Gardner, and R. Holzman. "Effect of Air Pollution on Alteration of Susceptibility to Pulmonary Infection," in *Proceedings of the Third Annual Conference on Atmospheric Contamination in Confined Spaces*, May, 1967, Dayton, OH. (Wright-Patterson Air Force Base, OH: Aerospace Medical Research Laboratories, 1967), pp. 71–80; report no. AMRL-TR-67-200. Available from: NTIS, Springfield, VA; AD-835008.

96. Ehrlich, R., J. C. Findlay, J. D. Fenters, and D. E. Gardner. "Health Effects of Short-Term Inhalation of Nitrogen Dioxide and Ozone Mixtures," *Environ. Res.* 14:223–231 (1977).

97. Miller, F. J., J. W. Illing, and D. E. Gardner. "Effect of Urban Ozone Levels on Laboratory-Induced Respiratory Infections," *Toxicol. Lett.* 2:163–169 (1978a).

98. Aranyi, C., S. C. Vana, P. T. Thomas, J. N. Bradof, J. D. Fenters, J. A. Graham, and F. J. Miller. "Effects of Subchronic Exposure to a Mixture of O_3, SO_2, and $(NH_4)_2SO_4$ on Host Defenses of Mice," *J. Toxicol. Environ. Health* 12:55–71 (1983).

99. Coffin, D. L., D. E. Gardner, R. S. Holzman, and F. J. Wolock. "Influence of Ozone on Pulmonary Cells," *Arch. Environ. Health* 16: 633–636 (1968).

100. Coffin, D. L., and D. E. Gardner. "Interaction of Biological Agents and Chemical Air Pollutants," *Ann. Occup. Hyg.* 15:219–235 (1972).

101. Goldstein, B. D., L. Y. Lai, and R. Cuzzi-Spada. "Potentiation of Complement-Dependent Membrane Damage by Ozone," *Arch. Environ. Health* 28:40–42 (1974).

102. Goldstein, B. D., S. J. Hamburger, G. W. Falk, and M. A. Amoruso. "Effect of Ozone and Nitrogen Dioxide on the Agglutination of Rat Alveolar Macrophages by Concanavalin A," *Life Sci.* 21:1637–1644 (1977).

103. Ehrlich, R., J. C. Findlay, and D. E. Gardner. "Effects of Repeated Exposures to Peak Concentrations of Nitrogen Dioxide and Ozone on Resistance to Streptococcal Pneumonia," *J. Toxicol. Environ. Health* 5:631–642 (1979).

104. Phalen, R. F., J. L. Kenoyer, T. T. Crocker, and T. R. McClure. "Effects of Sulfate Aerosols in Combination with Ozone on Elimination of Tracer Particles Inhaled by Rats," *J. Toxicol. Environ. Health* 6:797–810 (1980).

105. Frager, N. B., R. F. Phalen, and J. L. Kenoyer. "Adaptations to Ozone in Reference to Mucociliary Clearance," *Arch. Environ. Health* 34:51–57 (1979).

106. Kenoyer, J. L., R. F. Phalen, and J. R. Davis. "Particle Clearance from the Respiratory Tract as a Test of Toxicity: Effect of Ozone on Short and Long Term Clearance," *Exp. Lung Respir.* 2:111–120 (1981).

107. Abraham, W. M., A. J. Januszkiewicz, M. Mingle, M. Welker, A. Wanner, and M. A. Sackner. "Sensitivity of Bronchoprovocation and Tracheal Mucous Velocity in Detecting Airway Responses to O_3," *J. Appl. Physiol.: Respir. Environ. Exercise Physiol.* 48: 789–793 (1980).

108. Campbell, K. I., and R. H. Hilsenroth. "Impaired Resistance to Toxin in Toxoid-Immunized Mice Exposed to Ozone or Nitrogen Dioxide," *Clin. Toxicol.* 9:943–954 (1976).

109. Thomas, G. B., J. D. Fenters, R. Ehrlich, and D. E. Gardner. "Effects of Exposure to Ozone on Susceptibility to Experimental Tuberculosis," *Toxicol. Lett.* 9:11–17 (1981).

110. Fujimaki, H., M. Ozawa, T. Imai, and F. Shimizu. "Effect of Short-Term Exposure to O_3 on Antibody Response in Mice," *Environ. Res.* 35:490–496 (1984).

111. Alpert, S. M., D. E. Gardner, D. J. Hurst, T. R. Lewis, and D. L. Coffin. "Effects of Exposure to Ozone on Defensive Mechanisms of the Lung," *J. Appl. Physiol.* 31:247–252 (1971).

112. Dowell, A. R., L. A. Lohrbauer, D. Hurst, and S. D. Lee. "Rabbit Alveolar Macrophage Damage Caused by *in vivo* Ozone Inhalation," *Arch. Environ. Health* 21:121–127 (1970).

113. Hurst, D. J., D. E. Gardner, and D. L. Coffin. "Effect of Ozone on Acid Hydrolases of the Pulmonary Alveolar Macrophage," *J. Reticuloendothel. Soc.* 8:288–300 (1970).

114. Hurst, D. J., and D. L. Coffin. "Ozone Effect on Lysosomal Hydrolases of Alveolar Macrophages *in vitro*," *Arch. Intern. Med.* 127:1059–1063 (1971).

115. Goldstein, E., W. S. Tyler, P. D. Hoeprich, and C. Eagle. Ozone and the Antibacterial Defense Mechanisms of the Murine Lung," *Arch. Intern. Med.* 128:1099–1102 (1971a).

116. Goldstein, E., W. S. Tyler, P. D. Hoeprich, and C. Eagle. "Adverse Influence of Ozone on Pulmonary Bactericidal Activity of Murine Lungs," *Nature* (London) 229:262–263 (1971b).

117. Hadley, J. G., D. E. Gardner, D. L. Coffin, and D. B. Menzel. "Enhanced Binding of Autologous Cells to the Macrophage plasma Membrane as a Sensitive Indicator of Pollutant Damage," in C. L. Sanders, R. P. Schneider, G. E. Dagle and H. A. Ragan, eds. *Pulmonary Macrophage and Epithelial Cells: Proceedings of the Sixteenth Annual Hanford Biology Symposium,* September 1976, Richland, WA. (Washington, DC: Energy Research and Development Administration, 1977), pp. 1–21. (ERDA Symposium Series: v. 43). Available from: NTIS, Springfield, VA; CONF-760927.

118. McAllen, S. J., S. P. Chiu, R. F. Phalen, and R. E. Rasmussen. "Effect of *in vivo* Ozone Exposure on *in vitro* Pulmonary Alveolar Macrophage Mobility," *J. Toxicol. Environ. Health* 7:373–381 (1981).

119. Witz, G., M. A. Amoruso, and B. D. Goldstein. "Effect of Ozone on Alveolar Macrophage Function: Membrane Dynamic Properties," in S. D. Lee, M. G. Mustafa, and M. A. Mehlman, eds. *International Symposium on the Biomedical Effects of Ozone and Related Photochemical Oxidants,* March 1982, Pinehurst, NC. (Princeton, NJ: Princeton Scientific Publishers, Inc., 1983), pp. 263–272. (Advances in Modern Environmental Toxicology: v. 5).

120. Amoruso, M. A., G. Witz, and B. D. Goldstein. Decreased Superoxide Anion Radical Production by Rat Alveolar Macrophages Following Inhalation of Ozone or Nitrogen Dioxide," *Life Sci.* 28: 2215–2221 (1981).

121. Gardner, D. E., F. J. Miller, J. W. Illing, and J. M. Kirtz. "Increased Infectivity with Exposure to Ozone and Sulfuric Acid," *Toxicol. Lett.* 1:59–64 (1977).

122. Ehrlich, R. "Interaction Between Environmental Pollutants and Respiratory Infections," *EHP Environ. Health Perspect.* 35: 89–100 (1980).
123. Grose, E. C., D. E. Gardner, and F. J. Miller. "Response of Ciliated Epithelium to Ozone and Sulfuric Acid," *Environ. Res.* 22: 377–385 (1980).
124. Grose, E. C., J. H. Richards, J. W. Illing, F. J. Miller, D. W. Davies, J. A. Graham, and D. E. Gardner. "Pulmonary Host Defense Responses to Inhalation of Sulfuric Acid and Ozone," *J. Toxicol. Environ. Health* 10:351–362 (1982).
125. Illing, J. W., F. J. Miller, and D. E. Gardner. "Decreased Resistance to Infection in Exercised Mice Exposed to NO_2 and O_3," *J. Toxicol. Environ. Health* 6:843–851 (1980).
126. Green, G. M. "Similarities of Host Defense Mechanisms Against Pulmonary Disease in Animals and Man," *J. Toxicol. Environ. Health* 13:471–478 (1984).
127. Miller, F. J., J. H. Overton, Jr., R. H. Jaskot, and D. B. Menzel. "A Model of the Regional Uptake of Gaseous Pollutants in the Lung. I. The Sensitivity of the Uptake of Ozone in the Human Lung to Lower Respiratory Tract Secretions and to Exercise," *Toxicol. Appl. Pharmacol.* 79:11–27 (1985).
128. Guerrero, R. R., D. E. Rounds, R. S. Olson, and J. D. Hackney. "Mutagenic Effects of Ozone on Human Cells Exposed *in vivo* and *in vitro* Based on Sister Chromatid Exchange Analysis," *Environ. Res.* 18:336–346 (1979).
129. Zelac, R. E., H. L. Cromroy, W. E. Bolch, Jr., B. G. Dunavant, and H. A. Bevis. "Inhaled Ozone as a Mutagen. I. Chromosome Aberrations Induced in Chinese Hamster Lymphocytes," *Environ. Res.* 4:262–282 (1971a).
130. Zelac, R. E., H. L. Cromroy, W. E. Bolch, Jr., B. G. Dunavant, and H. A. Bevis. "Inhaled Ozone as a Mutagen. II. Effect on the Frequency of Chromosome Aberrations Observed in Irradiated Chinese Hamsters," *Environ. Res.* 4:325–342 (1971b).
131. Tice, R. R., M. A. Bender, J. L. Ivett, and R. T. Drew. "Cytogenetic Effects of Inhaled Ozone," *Mutat. Res.* 58:293–304 (1978).
132. Gooch, P. C., D. A. Creasia, and J. G. Brewen. "The Cytogenetic Effect of Ozone: Inhalation and *in vitro* Exposures," *Environ. Res.* 12:188–195 (1976).
133. Merz, T., M. A. Bender, H. D. Kerr, and T. J. Kulle. "Observations of Aberrations in Chromosomes of Lymphocytes from Human Subjects Exposed to Ozone at a Concentration of 0.5 ppm for 6 and 10 Hours," *Mutat. Res.* 31:299–302 (1975).
134. McKenzie, W. H., J. H. Knelson, N. J. Rummo, and D. E. House. "Cytogenetic Effects of Inhaled Ozone in Man," *Mutat. Res.* 48: 95–102 (1977).
135. McKenzie, W. H. "Controlled Human Exposure Studies: Cytogenetic Effects of Ozone Inhalation," in B. A. Bridges, B. E. Butterworth, and I. B. Weinstein, eds. *Indicators of Genotoxic Exposure.* (Spring Harbor, NY: Cold Spring Harbor Laboratory, 1982), pp. 319–324. (Banbury Report: no. 13).
136. Magie, A. R., D. E. Abbey, and W. R. Centerwall. "Effect of Photochemical Smog on the Peripheral Lymphocytes of Nonsmoking College Students," in *Environ. Res.* 29:204–219 (1982).

CHAPTER 12

EPA's Implementation Strategy for Ozone Reductions

John Hanisch

Large areas of the nation are still experiencing violations of the ozone standard.
Before discussing the Environmental Protection Agency's post-1987 ozone strategy, we should discuss what the EPA required for an approved 1982 Ozone State Implementation Plan (SIP), why the strategy did not work, and finally, what is available for future reductions.

What EPA Required for an Approved 1982 Ozone SIP

The 1982 Ozone SIP policy required states that could not attain the ambient ozone standard by December 31, 1982 to develop a SIP with the following basic elements:

1. an emissions inventory
2. an inspection and maintenance program (I/M)
3. stationary source controls (CTGs)
4. transportation control measures (RACMs)
5. a modeling demonstration using EKMA or other EPA approved photo-chemical model
6. an attainment date no later than December 31, 1987

Each of these measures is discussed below:

1. Emissions inventory: The states were required to identify existing emissions from mobile, stationary, and area sources. Using population projections consistent

with national guidance, the states also projected emissions with and without controls for 1987.

2. Inspection and maintenance: The I/M program is an in-use vehicle program which includes a test of emissions from cars in the idle mode. It may also include an antitampering check to see that the gas-fill restrictor and other engine components are in place.

3. Stationary source controls: The stationary source controls were divided into four groups. Group I was required with the 1979 SIPs and included controls on sources such as on coaters, painters, and gasoline storage facilities. Group II was required with the 1982 SIP, and included controls on sources such as in metal parts coating, graphic arts, and gasoline trucks. Group III was required after the 1982 SIP was approved; it includes sources such as resins manufacturing, synthetic organic manufacturing, and petroleum dry cleaning. The EPA published guidance documents called "Control Technology Guidance" (CTGs) for states to use in developing regulations for all sources listed under Group I, II, and III. The CTGs identify the level of control that is considered "RACT," or Reasonable Available Control Technology for each source category. A complete list of all the CTG source categories can be found in Appendix A.

The Group IV sources are sources such as golf ball manufacturing, boat builders, and staple manufacturers. The EPA has not published guidance on RACT for these sources. It has been up to the states to evaluate the sources and determine RACT.

4. Mobile sources controls: Mobile source controls were a key element of the 1982 SIPs. Congress targeted money through the Metropolitan Planning Organization (MPOs) to study transportation measures which could be implemented to reduce automobile emissions through reduction in Vehicle Miles of Travel (VMT). The MOs considered measures such as:

parking freeze
vehicle lanes
toll free structure
parking fee structure

auto force zone high occupancy
right turn on red
fringe parking

Each 1982 SIP was to include those measures which were Reasonably Available Control Measures (RACMs).

5. Modeling: Finally, the 1982 SIPs were to include modeling analysis using city-specific EKMA; this modeling analysis was designed to determine the amount of VOC reduction needed to bring the area into attainment by, or before, December 31, 1987. Additionally, for the east coast, the states were to participate in and develop data for the Northeast Corridor Regional Modeling Project. The EPA and the states spent $12 million developing databases and measuring ambient air; included in this was upper atmosphere sampling on high ozone days.

We still have multiple exceedences of the standard at many sites in Connecticut, Rhode Island, Massachusetts, and Maine. The sites upwind in Maryland,

Pennsylvania, New Jersey, and New York are also still experiencing very high ozone levels. This, despite the fact that many of these states say they are achieving the emission reduction targeted in their SIPs. Why are the ozone levels so high?

First, let's look at the attainment demonstration. The states were told they should assume that the air that would be coming into the nonattainment area would be at, or below, the standard. This has not happened. If the air coming in is above the standard, it is impossible to reach attainment.

Although the states have achieved their projected VOC emission reductions, it is clear that they have not achieved all of the reductions they could have achieved. For example, the states were allowed to build growth into their emission inventories. As a result, although the general trend in VOC emissions has been down, there have been increases in VMT, increases in industrial activity, and increases in various area-wide sources.

The increases in VMT, coupled with the fact that the Reid Vapor pressure of gasoline being sold (11 psi vs the 9 psi used to certify new cars) has caused more emissions from in-use vehicles than initially projected in the states' SIPs.

We have not achieved everything we should from stationary source controls either. In the stationary source area there are two distinct problems. The first is that guidance the EPA gave the states did not contain recordkeeping and reporting requirements, or accurate test methods. As a result, we presently do not have good data on how many sources are in compliance with the federal requirements. Additionally, many states developed regulations with size cut-offs, emission limits that are not as stringent as recommended, and averaging times which are inconsistent with solving violations of a one-hour ozone standard.

A third serious problem is that degreasing sources, gas stations, trucks that deliver gasoline to gas stations, and dry cleaning sources are all sources where there are many individual small emitters. The states do not have the resources to inspect all of these sources, and the regulations are written so that the state must enforce against the individual source. (The degreasing regulations requires every degreaser to be closed when not in use. It does not prohibit the sale or use of degreasers without automatically closing covers.)

The success stories in Massachusetts and Connecticut have been the Federal Motor Vehicle Emission Control Program (FMVECP), I/M, and Stage I controls for gasoline stations. In 1985, these three programs were responsible for 92% of the total emissions reduction achieved by Connecticut. The FMVECP achieved a 98,000 kg/day reduction. The I/M program achieved a 12,600 kg/day reduction, and Stage I achieved an 8,300 kg/day reduction.

What Is Available for Future Reductions

Everything that is relatively easy has been done. What we now need to do is increase the geographic area of coverage, get rid of the deviations that exist in the existing regulations, aggressively enforce against sources not complying with existing regulations, and look to area type sources for more controls.

There are some area type sources that have not been controlled. Included are sources such as auto refinishing; treatment storage and disposal facilities; treatment works, both publicly and privately owned; consumer/commercial solvents; paints; and Stage II. A complete list is provided in Appendix B. The Administrator is studying these control options now. He is also evaluating whether to require that the Reid Vapor Pressure of gasoline be restricted to 9 pounds per square inch and the feasibility of future on-board controls for new cars.

On the east coast where the urbanized areas are close to one another, and transport of ozone and VOCs are a major problem, it is especially important to control a larger area, and the agency has committed to run a regional model (ROM) to determine what needs to be done to solve the problem. The results from the model will not be available for several years. In the interim, to protect the health of the public, and our plants and trees, we need to implement control in the areas listed above. After the results of the modeling are available we can identify additional reductions, hopefully in isolated areas, which will be needed to bring the entire area into attainment.

APPENDIX A
CONTROL TECHNIQUES GUIDELINE DOCUMENTS

Group 1

Gasoline Loading Terminals
Gasoline Bulk Plants
Service Stations—Stage I
Fixed Roof Petroleum Tanks
Miscellaneous Refinery Sources
Cutback Asphalt
Solvent Metal Cleaning

Surface Coating of:
  Cans
  Metal Coils
  Fabrics
  Paper Products
  Automobiles and Light Duty Trucks
  Magnetic Wire
  Magnetic Wire
  Large Appliances

Group II

Leaks from Petroleum Refineries
Miscellaneous Metal Parts Surface Coating
Surface Coating of Flat Wood Paneling
Synthetic Pharmaceutical Manufacture
Rubber Tire Manufacture
External Floating Roof Petroleum Tanks Graphic Arts
Perchloroethylene Dry Cleaning
Gasoline Truck Leaks and Vapor Collection

Group III

Manufacture of High-Density Polyethylene, Polypropylene, and Polystyrene Resins
Fugitive Emissions from Synthetic Organic Chemical, Polymer and Resin
    Manufacturing Equipment
Large Petroleum Dry Cleaners
Air Oxidation Processes—Synthetic Organic Chemical Manufacturing Industries
Equipment Leaks from Natural Gas/Gasoline Processing Plants

APPENDIX B
POTENTIAL STATIONARY SOURCE FOR POST 1987 CONTROL

Wood Furniture Coating
Autobody Refinishing
Metal Rolling
SOCMI Distillation
SOCMI Batch Process
Coke Oven By-Product Plants
Treatment Storage and Disposal Facilities
Publicly Owned Treatment Works
Privately Owned Treatment Works
Bakeries
WEB Offset Lithography
Plastics Parts Coating
Oil and Gas Production
Marine Vessels—Including Off-Loading Fuel
Stage II Vapor Recovery
Architectural Coatings
Commercial Solvents
Consumer Solvents
Traffic Paint

List of Contributors

Adams, Mary Beth; U.S.D.A. Forest Service; 5 Radnor Corporate Center; Suite 200; 100 Matsonford Road; Radnor, PA 19087

Barry, Brenda E; Health Effects Institute; 215 First Street; Cambridge, MA 02142

Beck, Barbara D., Gradient Corporation, 44 Brattle Street, Cambridge, MA 02138

Burkhart, Richard P.; Region 1 U.S. EPA; Environmental Services Division; 60 Westview Street; Lexington, MA 02173

Calabrese, Edward J.; Division of Public Health; University of Massachusetts; Amherst, MA 01003

Canada, Andrew T.; P.O. Box 3094; Department of Anesthesiology; Duke University Medical Center; Durham, NC 27710

Demerjian, Kenneth L.; Atmospheric Sciences Research Center; State University of New York; 100 Fuller Road; Albany, NY 12205

Dockery, Douglas W.; Harvard School of Public Health, Environmental Epidemiology Program; 665 Huntington Avenue; Boston, MA 02115

Gilbert, Charles E.; Division of Public Health; University of Massachusetts; Amherst, MA 01003

Hanisch, John; Region 1 U.S. EPA; John F. Kennedy Building; Boston, MA 02203

Kriebel, David; Department of Work Environment; University of Lowell; 1 University Avenue; Lowell, MA 01854

Kuczkowski, Joseph A.; Department 400A; Chemical & Specialty Polymers R&D; The Goodyear Tire & Rubber Company; 142 Goodyear Boulevard; Akron, OH 44305

Manning, William J.; Department of Plant Pathology; Fernald Hall; University of Massachusetts; Amherst, MA 01003

McDonnell, William; Medical Research Building C-224H; CB-7315; University of North Carolina; Chapel Hill, NC 27514

McKee, David J.; MD-12 OAQPS; U.S. Environmental Protection Agency; Research Triangle Park, NC 27711

Sham, Chi Ho; Center for Energy and Environmental Studies; Boston University; 675 Commonwealth Avenue; Boston, MA 02215

Taylor, George Evans, Jr.; Desert Research Institute; P.O. Box 60220; Reno, NV 89506

Index

acetylcholinesterase, 134
acid phosphatase, 106
acute crop injury, 58–60
acute exposure, 117, 128, 146, 164
acute response attenuation, 129
acylperoxyl radicals, 5
agglutination decrement, 134
air standards review process, 163
airway constriction, 108
airway inflammation, 129
airway obstructive disease, 177
airway permeability, 119
alcohols, 13
aldehydes, 4, 8, 11–13, 93, 175
alfalfa (*Medicago sativa* L.), 61
alkanes, 4, 9, 11, 13
alkenes, 4, 11, 13
alkylperoxyl radicals, 5
alveolar epithelium, 118
alveolar permeability, 119
alveolar septa, 110–111
ambient ozone, 148, 153, 157, 168
American elm (*Ulmus americana* L.), 79
American Thoracic Society, 177
animal age variable, 107
animal mortality studies, 179
animal toxicology studies, 166, 174
antioxidant protection of rubber, 95
antioxidants, 1–6, 97
antiozonants, 94, 96–99, 101
apnea, 139
aromatics, 4, 11, 13
asthma, 147, 153–55, 157, 168, 176–77
atmospheric ozone. *See* ozone formation chemistry
atmospheric photoxidation, 6
atmospheric sinks, 10
Atmoshpheric Trajectory and Diffusion Model (ATAD), 50
atomic oxygen [O_3P], 4

background ozone, 2
bacterial infections, 179
barrier films, 95
behavioral studies, 136
benomyl, 60–61
Bermuda high, 34
beta-glucuronidase, 106
blood absorption, 133
bradycardia, 139
Branching Atmospheric Trajectory Model (BAT), 50, 51
breathing frequency (f_B), 166
bronchioles, 109
bronchoconstriction, 118

camp studies, 148–150, 157
carbohydrate levels, trees, 79–80, 83
carbon allocation, 71, 78, 81, 85
carbon dioxide (CO_2) assimilation, 68, 70
carbon monoxide (CO), 34
carbon monoxide (CO) oxidation, 2
carbonyl oxide, 93
cardiovascular effects of exposure, 139
carnation (*Dianthus caryophyllus* L.), 60
catalese, 108
cell damage, active byproducts, 118
centriacinar region, 109, 178
cerebral catecholamine reduction, 137
chain repair theory, 101
chamber studies, 117, 147, 156
chemical kinetics, 3
chemical mechanism development, 2–3, 14
children and adolescents, exposure studies, 148–151, 157, 168, 175
cholinesterase, 140
chromosomal aberrations, 139
chronic bronchitis, 176
chronic crop injury, 58–60
chronic exposure, 164, 174